An Introduction to Earth's Atmosphere

An Introduction to Earth's Atmosphere

Editor: Jacob Goddard

R CALLISTO
REFERENCE

www.callistoreference.com

Callisto Reference,
118-35 Queens Blvd., Suite 400,
Forest Hills, NY 11375, USA

Visit us on the World Wide Web at:
www.callistoreference.com

ISBN: 978-1-64116-008-7 (Hardback)

Trademark Notice: Registered trademark of products or corporate names are used only for explanation and identification without intent to infringe.

Cataloging-in-Publication Data

An introduction to earth's atmosphere / edited by Jacob Goddard.
    p. cm.
Includes bibliographical references and index.
ISBN 978-1-64116-008-7
1. Atmosphere. 2. Earth (Planet). 3. Meteorology. I. Goddard, Jacob.
QC861.3 .I58 2018
551.5--dc23

Table of Contents

Permissions

Index

Preface

Earth's atmosphere consists of different gases, namely nitrogen, oxygen, argon and carbon dioxide with water vapors. The study of air is called aerology. The various layers in Earth's atmosphere are thermosphere, stratosphere, exosphere, troposphere and mesosphere. This book is a valuable compilation of topics, ranging from the basic to the most complex theories and principles in the study of Earth's atmosphere. The topics included in it are of utmost significance and bound to provide incredible insights to readers. Coherent flow of topics, student-friendly language and extensive use of examples make this textbook an invaluable source of knowledge.

A detailed account of the significant topics covered in this book is provided below:

Chapter 1- The atmosphere of the Earth surrounds it and preserves it by absorbing dangerous ultraviolet solar radiation. Some of the subjects that study the atmosphere are atmospheric sciences and atmospheric physics. This is an introductory chapter that will introduce briefly all the significant aspects of the Earth's atmosphere.

Chapter 2- The Earth's atmosphere can be divided into several layers. The five layers that exist are troposphere, stratosphere, mesosphere, thermosphere and exosphere. Tropopause is the boundary between the troposphere and the stratosphere. The major components of the Earth's atmosphere are discussed in this chapter.

Chapter 3- The branch of atmospheric science that studies the chemical composition of the atmosphere of the Earth is known as atmospheric chemistry. The subject focuses on the changes faced by the atmosphere because of global warming. Some of the harms studied are ozone depletion, greenhouse gases, acid rain, etc. Atmospheric chemistry is best understood in confluence with the major topics listed in the following chapter.

Chapter 4- The physical properties of the Earth's atmosphere discussed in the section are atmospheric pressure, atmospheric temperature, sunlight and density of air. Atmospheric pressure is the pressure at any given point in the Earth's atmosphere whereas atmospheric temperature is the temperature of the atmosphere decided by phenomena like humidity and altitude. The chapter serves as a source to understand the major categories related to the atmosphere of Earth.

Chapter 5- The atmospheric circulation of our planet differs every year. Polar vortex, Walker circulation and El Niño–Southern Oscillation are some of the topics important to the subject of atmospheric circulation. The major components of atmospheric circulation are discussed in this chapter.

Chapter 6- The thermodynamic effects that occur in the Earth's atmosphere and causes atmospheric phenomena such as weather and climate are studied under atmospheric thermodynamics. Related topics such as atmospheric convection, atmospheric instability, atmospheric sounding have been included in this section. The aspects elucidated in this chapter are of vital importance, and provide a better understanding of the Earth's atmosphere.

Chapter 7- Atmospheric temperature is measured by using various tools, techniques and concepts. The scale of temperature quantitatively measures temperature through the ideal gas scale and the International temperature scale of 1990. The chapter strategically encompasses and incorporates the major components and key concepts of atmospheric temperature measurement, providing a complete understanding.

It gives me an immense pleasure to thank our entire team for their efforts. Finally in the end, I would like to thank my family and colleagues who have been a great source of inspiration and support.

Editor

1

An Introduction to Atmosphere of the Earth

The atmosphere of the Earth surrounds it and preserves it by absorbing dangerous ultraviolet solar radiation. Some of the subjects that study the atmosphere are atmospheric sciences and atmospheric physics. This is an introductory chapter that will introduce briefly all the significant aspects of the Earth's atmosphere.

Atmosphere of Earth

Blue light is scattered more than other wavelengths by the gases in the atmosphere, giving Earth a blue halo when seen from space onboard *ISS* at a height of 402–424 km (250–263 mi).

The atmosphere of Earth is the layer of gases, commonly known as air, that surrounds the planet Earth and is retained by Earth's gravity. The atmosphere of Earth protects life on Earth by absorbing ultraviolet solar radiation, warming the surface through heat retention (greenhouse effect), and reducing temperature extremes between day and night (the diurnal temperature variation).

By volume, dry air contains 78.09% nitrogen, 20.95% oxygen, 0.93% argon, 0.04% carbon dioxide, and small amounts of other gases. Air also contains a variable amount of water vapor, on average around 1% at sea level, and 0.4% over the entire atmosphere. Air content and atmospheric pressure vary at different layers, and air suitable for use in photosynthesis by terrestrial plants and breathing of terrestrial animals is found only in Earth's troposphere and in artificial atmospheres.

The atmosphere has a mass of about 5.15×10^{18} kg, three quarters of which is within about 11 km (6.8 mi; 36,000 ft) of the surface. The atmosphere becomes thinner and thinner with increasing altitude, with no definite boundary between the atmosphere and outer space. The Kármán line, at 100 km (62 mi), or 1.57% of Earth's radius, is often used as the border between the atmosphere and outer space. Atmospheric effects become noticeable during atmospheric reentry of spacecraft at an altitude of around 120 km (75 mi). Several layers can be distinguished in the atmosphere, based on characteristics such as temperature and composition.

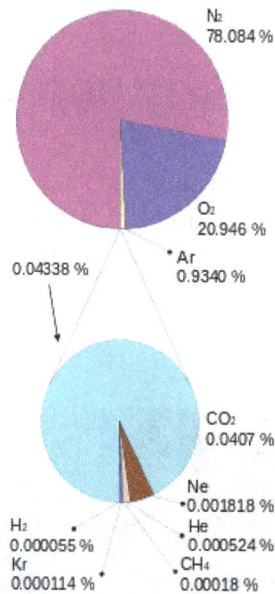

Composition of Earth's atmosphere by volume. The lower pie represents the trace gases that together compose about 0.038% of the atmosphere (0.043% with CO_2 at 2014 concentration). The numbers are from a variety of years (mainly 1987, with CO_2 and methane from 2009) and do not represent any single source.

Diagram of Earth's atmosphere (layers to scale). Distance from the surface to the top of the stratosphere is just under 1% of Earth's radius.

The study of Earth's atmosphere and its processes is called atmospheric science (aerology). Early pioneers in the field include Léon Teisserenc de Bort and Richard Assmann.

Composition

The three major constituents of air, and therefore of Earth's atmosphere, are nitrogen, oxygen, and argon. Water vapor accounts for roughly 0.25% of the atmosphere by mass. The concentration of water vapor (a greenhouse gas) varies significantly from around 10 ppm by volume in the coldest portions of the atmosphere to as much as 5% by volume in hot, humid air masses, and concentrations of other atmospheric gases are typically quoted in terms of dry air (without water vapor). The remaining gases are often referred to as trace gases, among which are the greenhouse gases, principally carbon dioxide, methane, nitrous oxide, and ozone. Filtered air includes trace amounts of many other chemical compounds. Many substances of natural origin may be present in locally and seasonally variable small amounts as aerosols in an unfiltered air sample, including dust of mineral and organic composition, pollen and spores, sea spray, and volcanic ash. Various industrial pollutants also may be present as gases or aerosols, such as chlorine (elemental or in compounds), fluorine compounds and elemental mercury vapor. Sulfur compounds such as hydrogen sulfide and sulfur dioxide (SO_2) may be derived from natural sources or from industrial air pollution.

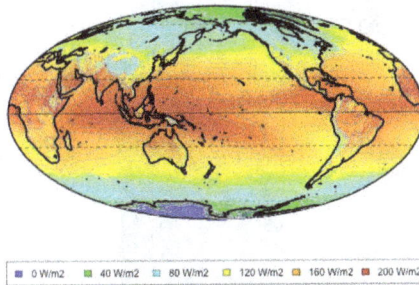

Distribution of Water Vapor
as shown by Clear-Sky Atmospheric Absorption
Avg Globe: 133 NH: 136.7 SH: 129.3 Trop: 171.3
Arc: 60.5 Ant: 16.6 Land: 118.5 Ocean: 138.6 W/m2

0 W/m2   40 W/m2   80 W/m2   120 W/m2   160 W/m2   200 W/m2

Mean atmospheric water vapor

Major constituents of dry air, by volume			
Gas		Volume[A]	
Name	Formula	in ppmv[B]	in %
Nitrogen	N_2	780,840	78.084
Oxygen	O_2	209,460	20.946
Argon	Ar	9,340	0.9340
Carbon dioxide	CO_2	400	0.04
Neon	Ne	18.18	0.001818
Helium	He	5.24	0.000524
Methane	CH_4	1.79	0.000179
Not included in above dry atmosphere:			
Water vapor[C]	H_2O	10–50,000[D]	0.001%–5%[D]

notes:
(A)   volume fraction is equal to mole fraction for ideal gas only
(B)   ppmv: parts per million by volume
(C)   Water vapor is about 0.25% *by mass* over full atmosphere
(D)   Water vapor strongly varies locally

Structure of the Atmosphere

Principal Layers

In general, air pressure and density decrease with altitude in the atmosphere. However, temperature has a more complicated profile with altitude, and may remain relatively constant or even increase with altitude in some regions. Because the general pattern of the temperature/altitude profile is constant and measurable by means of instrumented balloon soundings, the temperature behavior provides a useful metric to distinguish atmospheric layers. In this way, Earth's atmosphere can be divided (called atmospheric stratification) into five main layers. Excluding the exosphere, the atmosphere has four primary layers, which are the troposphere, stratosphere, mesosphere, and thermosphere. From highest to lowest, the five main layers are:

- Exosphere: 700 to 10,000 km (440 to 6,200 miles)

- Thermosphere: 80 to 700 km (50 to 440 miles)

- Mesosphere: 50 to 80 km (31 to 50 miles)

- Stratosphere: 12 to 50 km (7 to 31 miles)

- Troposphere: 0 to 12 km (0 to 7 miles)

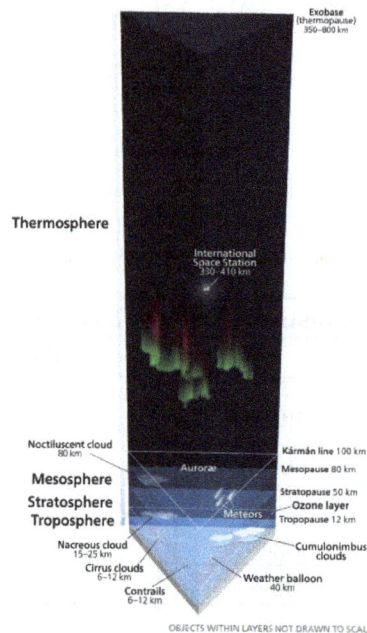

Earth's atmosphere Lower 4 layers of the atmosphere in 3 dimensions as seen diagonally from above the exobase. Layers drawn to scale, objects within the layers are not to scale. Aurorae shown here at the bottom of the thermosphere can actually form at any altitude in this atmospheric layer

Exosphere

The exosphere is the outermost layer of Earth's atmosphere (i.e. the upper limit of the atmosphere). It extends from the exobase, which is located at the top of the thermosphere at an altitude of about 700 km above sea level, to about 10,000 km (6,200 mi; 33,000,000 ft) where it merges into the solar wind.

This layer is mainly composed of extremely low densities of hydrogen, helium and several heavier molecules including nitrogen, oxygen and carbon dioxide closer to the exobase. The atoms and molecules are so far apart that they can travel hundreds of kilometers without colliding with one another. Thus, the exosphere no longer behaves like a gas, and the particles constantly escape into space. These free-moving particles follow ballistic trajectories and may migrate in and out of the magnetosphere or the solar wind.

The exosphere is located too far above Earth for any meteorological phenomena to be possible. However, the aurora borealis and aurora australis sometimes occur in the lower part of the exosphere, where they overlap into the thermosphere. The exosphere contains most of the satellites orbiting Earth.

Thermosphere

The thermosphere is the second-highest layer of Earth's atmosphere. It extends from the mesopause (which separates it from the mesosphere) at an altitude of about 80 km (50 mi; 260,000 ft) up to the thermopause at an altitude range of 500–1000 km (310–620 mi; 1,600,000–3,300,000 ft). The height of the thermopause varies considerably due to changes in solar activity. Because the thermopause lies at the lower boundary of the exosphere, it is also referred to as the exobase. The lower part of the thermosphere, from 80 to 550 kilometres (50 to 342 mi) above Earth's surface, contains the ionosphere.

The temperature of the thermosphere gradually increases with height. Unlike the stratosphere beneath it, wherein a temperature inversion is due to the absorption of radiation by ozone, the inversion in the thermosphere occurs due to the extremely low density of its molecules. The temperature of this layer can rise as high as 1500°C (2700°F), though the gas molecules are so far apart that its temperature in the usual sense is not very meaningful. The air is so rarefied that an individual molecule (of oxygen, for example) travels an average of 1 kilometre (0.62 mi; 3300 ft) between collisions with other molecules. Although the thermosphere has a high proportion of molecules with high energy, it would not feel hot to a human in direct contact, because its density is too low to conduct a significant amount of energy to or from the skin.

This layer is completely cloudless and free of water vapor. However, non-hydrometeorological phenomena such as the aurora borealis and aurora australis are occasionally seen in the thermosphere. The International Space Station orbits in this layer, between 350 and 420 km (220 and 260 mi).

Mesosphere

The mesosphere is the third highest layer of Earth's atmosphere, occupying the region above the stratosphere and below the thermosphere. It extends from the stratopause at an altitude of about 50 km (31 mi; 160,000 ft) to the mesopause at 80–85 km (50–53 mi; 260,000–280,000 ft) above sea level.

Temperatures drop with increasing altitude to the mesopause that marks the top of this middle layer of the atmosphere. It is the coldest place on Earth and has an average temperature around −85°C (−120°F; 190 K).

Just below the mesopause, the air is so cold that even the very scarce water vapor at this altitude can be sublimated into polar-mesospheric noctilucent clouds. These are the highest clouds in the atmosphere and may be visible to the naked eye if sunlight reflects off them about an hour or two after sunset or a similar length of time before sunrise. They are most readily visible when the Sun is around 4 to 16 degrees below the horizon. A type of lightning referred to as either sprites or ELVES occasionally forms far above tropospheric thunderclouds. The mesosphere is also the layer where most meteors burn up upon atmospheric entrance. It is too high above Earth to be accessible to jet-powered aircraft and balloons, and too low to permit orbital spacecraft. The mesosphere is mainly accessed by sounding rockets and rocket-powered aircraft.

Stratosphere

The stratosphere is the second-lowest layer of Earth's atmosphere. It lies above the troposphere and is separated from it by the tropopause. This layer extends from the top of the troposphere at roughly 12 km (7.5 mi; 39,000 ft) above Earth's surface to the stratopause at an altitude of about 50 to 55 km (31 to 34 mi; 164,000 to 180,000 ft).

The atmospheric pressure at the top of the stratosphere is roughly 1/1000 the pressure at sea level. It contains the ozone layer, which is the part of Earth's atmosphere that contains relatively high concentrations of that gas. The stratosphere defines a layer in which temperatures rise with increasing altitude. This rise in temperature is caused by the absorption of ultraviolet radiation (UV) radiation from the Sun by the ozone layer, which restricts turbulence and mixing. Although the temperature may be −60°C (−76°F; 210 K) at the tropopause, the top of the stratosphere is much warmer, and may be near 0°C.

The stratospheric temperature profile creates very stable atmospheric conditions, so the stratosphere lacks the weather-producing air turbulence that is so prevalent in the troposphere. Consequently, the stratosphere is almost completely free of clouds and other forms of weather. However, polar stratospheric or nacreous clouds are occasionally seen in the lower part of this layer of the atmosphere where the air is coldest. The stratosphere is the highest layer that can be accessed by jet-powered aircraft.

Troposphere

The troposphere is the lowest layer of Earth's atmosphere. It extends from Earth's surface to an average height of about 12 km, although this altitude actually varies from about 9 km (30,000 ft) at the poles to 17 km (56,000 ft) at the equator, with some variation due to weather. The troposphere is bounded above by the tropopause, a boundary marked in most places by a temperature inversion (i.e. a layer of relatively warm air above a colder one), and in others by a zone which is isothermal with height.

Although variations do occur, the temperature usually declines with increasing altitude in the troposphere because the troposphere is mostly heated through energy transfer from the surface. Thus, the lowest part of the troposphere (i.e. Earth's surface) is typically the warmest section of the troposphere. This promotes vertical mixing (hence the origin of its name in the Greek word *tropos*, meaning "turn"). The troposphere contains roughly 80% of the mass of Earth's atmosphere. The troposphere is denser than all its overlying atmospheric layers because a larger atmospheric weight sits on top of the troposphere and causes it to be most severely compressed. Fifty percent of the total mass of the atmosphere is located in the lower 5.6 km (18,000 ft) of the troposphere.

Nearly all atmospheric water vapor or moisture is found in the troposphere, so it is the layer where most of Earth's weather takes place. It has basically all the weather-associated cloud genus types generated by active wind circulation, although very tall cumulonimbus thunder clouds can penetrate the tropopause from below and rise into the lower part of the stratosphere. Most conventional aviation activity takes place in the troposphere, and it is the only layer that can be accessed by propeller-driven aircraft.

Space Shuttle *Endeavour* orbiting in the thermosphere. Because of the angle of the photo, it appears to straddle the stratosphere and mesosphere that actually lie more than 250 km below. The orange layer is the troposphere, which gives way to the whitish stratosphere and then the blue mesosphere.

Other Layers

Within the five principal layers that are largely determined by temperature, several secondary layers may be distinguished by other properties:

- The ozone layer is contained within the stratosphere. In this layer ozone concentrations are about 2 to 8 parts per million, which is much higher than in the lower atmosphere but still very small compared to the main components of the atmosphere. It is mainly located in the lower portion of the stratosphere from about 15–35 km (9.3–21.7 mi; 49,000–115,000 ft), though the thickness varies seasonally and geographically. About 90% of the ozone in Earth's atmosphere is contained in the stratosphere.

- The ionosphere is a region of the atmosphere that is ionized by solar radiation. It is responsible for auroras. During daytime hours, it stretches from 50 to 1,000 km (31 to 621 mi; 160,000 to 3,280,000 ft) and includes the mesosphere, thermosphere, and parts of the exosphere. However, ionization in the mesosphere largely ceases during the night, so auroras are normally seen only in the thermosphere and lower exosphere. The ionosphere forms the inner edge of the magnetosphere. It has practical importance because it influences, for example, radio propagation on Earth.

- The homosphere and heterosphere are defined by whether the atmospheric gases are well mixed. The surface-based homosphere includes the troposphere, stratosphere, mesosphere, and the lowest part of the thermosphere, where the chemical composition of the atmosphere does not depend on molecular weight because the gases are mixed by turbulence. This relatively homogeneous layer ends at the *turbopause* found at about 100 km (62 mi; 330,000 ft), the very edge of space itself as accepted by the FAI, which places it about 20 km (12 mi; 66,000 ft) above the mesopause.

Above this altitude lies the heterosphere, which includes the exosphere and most of the thermosphere. Here, the chemical composition varies with altitude. This is because the distance that particles can move without colliding with one another is large compared with the size of motions that cause mixing. This allows the gases to stratify by molecular weight, with the heavier ones, such as oxygen and nitrogen, present only near the bottom of the heterosphere. The upper part of the heterosphere is composed almost completely of hydrogen, the lightest element.

- The planetary boundary layer is the part of the troposphere that is closest to Earth's surface and is directly affected by it, mainly through turbulent diffusion. During the day the planetary boundary layer usually is well-mixed, whereas at night it becomes stably stratified with weak or intermittent mixing. The depth of the planetary boundary layer ranges from as little as about 100 metres (330 ft) on clear, calm nights to 3,000 m (9,800 ft) or more during the afternoon in dry regions.

The average temperature of the atmosphere at Earth's surface is 14°C (57°F; 287 K) or 15°C (59°F; 288 K), depending on the reference.

Physical Properties

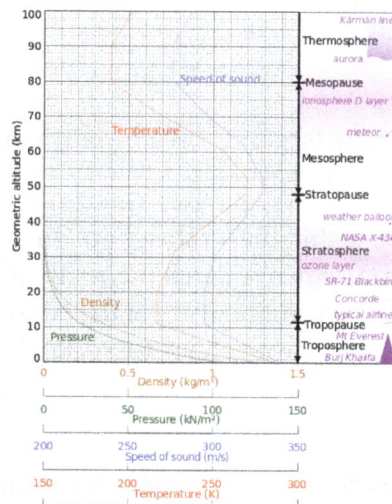

Comparison of the 1962 US Standard Atmosphere graph of geometric altitude against air density, pressure, the speed of sound and temperature with approximate altitudes of various objects.

Pressure and Thickness

The average atmospheric pressure at sea level is defined by the International Standard Atmosphere as 101325 pascals (760.00 Torr; 14.6959 psi; 760.00 mmHg). This is sometimes referred to as a unit of standard atmospheres (atm). Total atmospheric mass is 5.1480×10^{18} kg (1.135×10^{19} lb), about 2.5% less than would be inferred from the average sea level pressure and Earth's area of 51007.2 megahectares, this portion being displaced by Earth's mountainous terrain. Atmospheric pressure is the total weight of the air above unit area at the point where the pressure is measured.

Thus air pressure varies with location and weather.

If the entire mass of the atmosphere had a uniform density from sea level, it would terminate abruptly at an altitude of 8.50 km (27,900 ft). It actually decreases exponentially with altitude, dropping by half every 5.6 km (18,000 ft) or by a factor of 1/e every 7.64 km (25,100 ft), the average scale height of the atmosphere below 70 km (43 mi; 230,000 ft). However, the atmosphere is more accurately modeled with a customized equation for each layer that takes gradients of temperature, molecular composition, solar radiation and gravity into account.

In summary, the mass of Earth's atmosphere is distributed approximately as follows:

- 50% is below 5.6 km (18,000 ft).

- 90% is below 16 km (52,000 ft).

- 99.99997% is below 100 km (62 mi; 330,000 ft), the Kármán line. By international convention, this marks the beginning of space where human travelers are considered astronauts.

By comparison, the summit of Mt. Everest is at 8,848 m (29,029 ft); commercial airliners typically cruise between 10 km (33,000 ft) and 13 km (43,000 ft) where the thinner air improves fuel economy; weather balloons reach 30.4 km (100,000 ft) and above; and the highest X-15 flight in 1963 reached 108.0 km (354,300 ft).

Even above the Kármán line, significant atmospheric effects such as auroras still occur. Meteors begin to glow in this region, though the larger ones may not burn up until they penetrate more deeply. The various layers of Earth's ionosphere, important to HF radio propagation, begin below 100 km and extend beyond 500 km. By comparison, the International Space Station and Space Shuttle typically orbit at 350–400 km, within the F-layer of the ionosphere where they encounter enough atmospheric drag to require reboosts every few months. Depending on solar activity, satellites can experience noticeable atmospheric drag at altitudes as high as 700–800 km.

Temperature and Speed of Sound

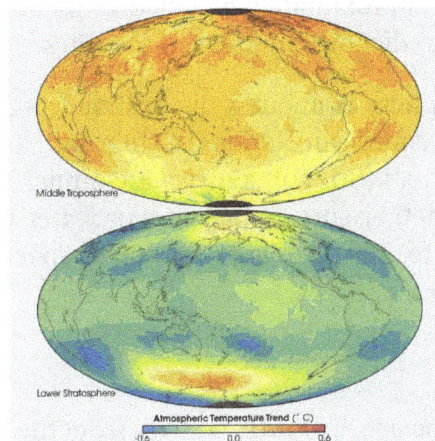

These images show temperature trends in two thick layers of the atmosphere as measured by a series of satellite-based instruments between January 1979 and December 2005. The measurements were taken by Microwave Sounding Units and Advanced Microwave Sounding Units flying on a series of National Oceanic and Atmospheric Administration (NOAA) weather satellites. The instruments record microwaves emitted from oxygen molecules in the atmosphere.

The division of the atmosphere into layers mostly by reference to temperature is discussed above. Temperature decreases with altitude starting at sea level, but variations in this trend begin above 11 km, where the temperature stabilizes through a large vertical distance through the rest of the troposphere. In the stratosphere, starting above about 20 km, the temperature increases with height, due to heating within the ozone layer caused by capture of significant ultraviolet radiation from the Sun by the dioxygen and ozone gas in this region. Still another region of increasing temperature with altitude occurs at very high altitudes, in the aptly-named thermosphere above 90 km.

Because in an ideal gas of constant composition the speed of sound depends only on temperature and not on the gas pressure or density, the speed of sound in the atmosphere with altitude takes on the form of the complicated temperature profile, and does not mirror altitudinal changes in density or pressure.

Density and Mass

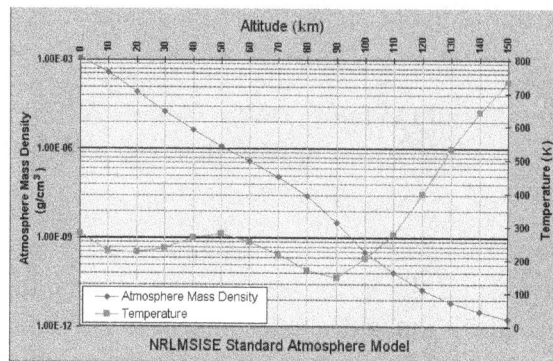

Temperature and mass density against altitude from the NRLMSISE-00 standard atmosphere model (the eight dotted lines in each "decade" are at the eight cubes 8, 27, 64, …, 729).

The density of air at sea level is about 1.2 kg/m³ (1.2 g/L, 0.0012 g/cm³). Density is not measured directly but is calculated from measurements of temperature, pressure and humidity using the equation of state for air (a form of the ideal gas law). Atmospheric density decreases as the altitude increases. This variation can be approximately modeled using the barometric formula. More sophisticated models are used to predict orbital decay of satellites.

The average mass of the atmosphere is about 5 quadrillion (5×10^{15}) tonnes or 1/1,200,000 the mass of Earth. According to the American National Center for Atmospheric Research, "The total mean mass of the atmosphere is 5.1480×10^{18} kg with an annual range due to water vapor of 1.2 or 1.5×10^{15} kg, depending on whether surface pressure or water vapor data are used; somewhat smaller than the previous estimate. The mean mass of water vapor is estimated as 1.27×10^{16} kg and the dry air mass as $5.1352 \pm0.0003\times10^{18}$ kg.

Optical Properties

Solar radiation (or sunlight) is the energy Earth receives from the Sun. Earth also emits radiation back into space, but at longer wavelengths that we cannot see. Part of the incoming and emitted radiation is absorbed or reflected by the atmosphere. In May 2017, glints of light, seen as twinkling from an orbiting satellite a million miles away, were found to be reflected light from ice crystals in the atmosphere.

Scattering

When light passes through Earth's atmosphere, photons interact with it through *scattering*. If the light does not interact with the atmosphere, it is called *direct radiation* and is what you see if you were to look directly at the Sun. *Indirect radiation* is light that has been scattered in the atmosphere. For example, on an overcast day when you cannot see your shadow there is no direct radiation reaching you, it has all been scattered. As another example, due to a phenomenon called Rayleigh scattering, shorter (blue) wavelengths scatter more easily than longer (red) wavelengths. This is why the sky looks blue; you are seeing scattered blue light. This is also why sunsets are red. Because the Sun is close to the horizon, the Sun's rays pass through more atmosphere than normal to reach your eye. Much of the blue light has been scattered out, leaving the red light in a sunset.

Absorption

Different molecules absorb different wavelengths of radiation. For example, O_2 and O_3 absorb almost all wavelengths shorter than 300 nanometers. Water (H_2O) absorbs many wavelengths above 700 nm. When a molecule absorbs a photon, it increases the energy of the molecule. This heats the atmosphere, but the atmosphere also cools by emitting radiation, as discussed below.

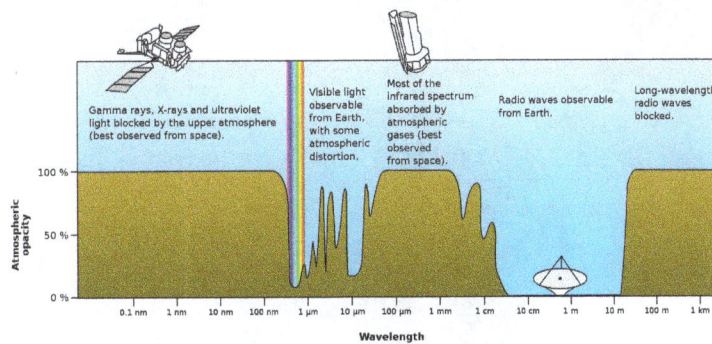

Rough plot of Earth's atmospheric transmittance (or opacity) to various wavelengths of electromagnetic radiation, including visible light.

The combined absorption spectra of the gases in the atmosphere leave "windows" of low opacity, allowing the transmission of only certain bands of light. The optical window runs from around 300 nm (ultraviolet-C) up into the range humans can see, the visible spectrum (commonly called light), at roughly 400–700 nm and continues to the infrared to around 1100 nm. There are also infrared and radio windows that transmit some infrared and radio waves at longer wavelengths. For example, the radio window runs from about one centimeter to about eleven-meter waves.

Emission

Emission is the opposite of absorption, it is when an object emits radiation. Objects tend to emit amounts and wavelengths of radiation depending on their "black body" emission curves, therefore hotter objects tend to emit more radiation, with shorter wavelengths. Colder objects emit less radiation, with longer wavelengths. For example, the Sun is approximately 6,000 K (5,730°C; 10,340°F), its radiation peaks near 500 nm, and is visible to the human eye. Earth is approximately 290 K (17°C; 62°F), so its radiation peaks near 10,000 nm, and is much too long to be visible to humans.

Because of its temperature, the atmosphere emits infrared radiation. For example, on clear nights Earth's surface cools down faster than on cloudy nights. This is because clouds (H_2O) are strong absorbers and emitters of infrared radiation. This is also why it becomes colder at night at higher elevations.

The greenhouse effect is directly related to this absorption and emission effect. Some gases in the atmosphere absorb and emit infrared radiation, but do not interact with sunlight in the visible spectrum. Common examples of these are CO_2 and H_2O.

Refractive Index

The refractive index of air is close to, but just greater than 1. Systematic variations in refractive index can lead to the bending of light rays over long optical paths. One example is that, under some circumstances, observers onboard ships can see other vessels just over the horizon because light is refracted in the same direction as the curvature of Earth's surface.

The refractive index of air depends on temperature, giving rise to refraction effects when the temperature gradient is large. An example of such effects is the mirage.

Circulation

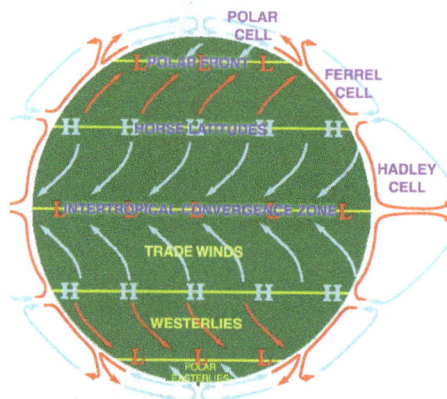

An idealised view of three large circulation cells.

Atmospheric circulation is the large-scale movement of air through the troposphere, and the means (with ocean circulation) by which heat is distributed around Earth. The large-scale structure of the atmospheric circulation varies from year to year, but the basic structure remains fairly constant because it is determined by Earth's rotation rate and the difference in solar radiation between the equator and poles.

Evolution of Earth's Atmosphere

Earliest Atmosphere

The first atmosphere consisted of gases in the solar nebula, primarily hydrogen. There were probably simple hydrides such as those now found in the gas giants (Jupiter and Saturn), notably water vapor, methane and ammonia. As the solar nebula dissipated, these gases escaped, partly driven off by the solar wind.

Second Atmosphere

Outgassing from volcanism, supplemented by gases produced during the late heavy bombardment of Earth by huge asteroids, produced the next atmosphere, consisting largely of nitrogen plus carbon dioxide and inert gases. A major part of carbon-dioxide emissions soon dissolved in water and built up carbonate sediments.

Researchers have found water-related sediments dating from as early as 3.8 billion years ago. About 3.4 billion years ago, nitrogen formed the major part of the then stable "second atmosphere". An influence of life has to be taken into account rather soon in the history of the atmosphere, because hints of early life-forms appear as early as 3.5 billion years ago. How Earth at that time maintained a climate warm enough for liquid water and life, if the early Sun put out 30% lower solar radiance than today, is a puzzle known as the "faint young Sun paradox".

The geological record however shows a continually relatively warm surface during the complete early temperature record of Earth - with the exception of one cold glacial phase about 2.4 billion years ago. In the late Archean Eon an oxygen-containing atmosphere began to develop, apparently produced by photosynthesizing cyanobacteria, which have been found as stromatolite fossils from 2.7 billion years ago. The early basic carbon isotopy (isotope ratio proportions) strongly suggests conditions similar to the current, and that the fundamental features of the carbon cycle became established as early as 4 billion years ago.

Ancient sediments in the Gabon dating from between about 2,150 and 2,080 million years ago provide a record of Earth's dynamic oxygenation evolution. These fluctuations in oxygenation were likely driven by the Lomagundi carbon isotope excursion.

Third Atmosphere

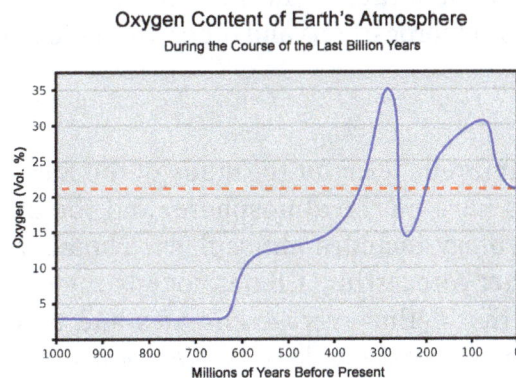

Oxygen content of the atmosphere over the last billion years.

The constant re-arrangement of continents by plate tectonics influences the long-term evolution of the atmosphere by transferring carbon dioxide to and from large continental carbonate stores. Free oxygen did not exist in the atmosphere until about 2.4 billion years ago during the Great Oxygenation Event and its appearance is indicated by the end of the banded iron formations. Before this time, any oxygen produced by photosynthesis was consumed by oxidation of reduced materials, notably iron. Molecules of free oxygen did not start to accumulate in the atmosphere until the rate of production of oxygen began to exceed the availability of reducing materials that removed oxygen. This point signifies a shift from a reducing atmosphere to an oxidizing atmosphere. O_2 showed major variations until reaching a steady state of more than 15% by the end of the Precambrian. The following time span from 541 million years ago to the present day is the Phanerozoic Eon, during the earliest period of which, the Cambrian, oxygen-requiring metazoan life forms began to appear.

The amount of oxygen in the atmosphere has fluctuated over the last 600 million years, reaching a peak of about 30% around 280 million years ago, significantly higher than today's 21%. Two main processes govern changes in the atmosphere: Plants use carbon dioxide from the atmosphere, releasing oxygen. Breakdown of pyrite and volcanic eruptions release sulfur into the atmosphere, which oxidizes and hence reduces the amount of oxygen in the atmosphere. However, volcanic eruptions also release carbon dioxide, which plants can convert to oxygen. The exact cause of the variation of the amount of oxygen in the atmosphere is not known. Periods with much oxygen in the atmosphere are associated with rapid development of animals. Today's atmosphere contains 21% oxygen, which is high enough for this rapid development of animals.

This image shows the buildup of tropospheric CO_2 in the Northern Hemisphere with a maximum around May. The maximum in the vegetation cycle follows, occurring in the late summer. Following the peak in vegetation, the drawdown of atmospheric CO_2 due to photosynthesis is apparent, particularly over the boreal forests.

The scientific consensus is that the anthropogenic greenhouse gases currently accumulating in the atmosphere are the main cause of global warming.

Air Pollution

Air pollution is the introduction into the atmosphere of chemicals, particulate matter or biological materials that cause harm or discomfort to organisms. Stratospheric ozone depletion is caused by air pollution, chiefly from chlorofluorocarbons and other ozone-depleting substances.

Atmospheric Sciences

Atmospheric sciences is an umbrella term for the study of the Earth's atmosphere, its processes, the effects other systems have on the atmosphere, and the effects of the atmosphere on these other systems. Meteorology includes atmospheric chemistry and atmospheric physics with a major focus on weather forecasting. Climatology is the study of atmospheric changes (both long and short-term) that define average climates and their change over time, due to both natural and anthropogenic climate variability. Aeronomy is the study of the upper layers of the atmosphere, where dissociation and ionization are important. Atmospheric science has been extended to the field of planetary science and the study of the atmospheres of the planets of the solar system.

Experimental instruments used in atmospheric sciences include satellites, rocketsondes, radiosondes, weather balloons, and lasers.

The term aerology (from Greek air and logia) is sometimes used as an alternative term for the study of Earth's atmosphere. Early pioneers in the field include Léon Teisserenc de Bort and Richard Assmann.

Atmospheric Chemistry

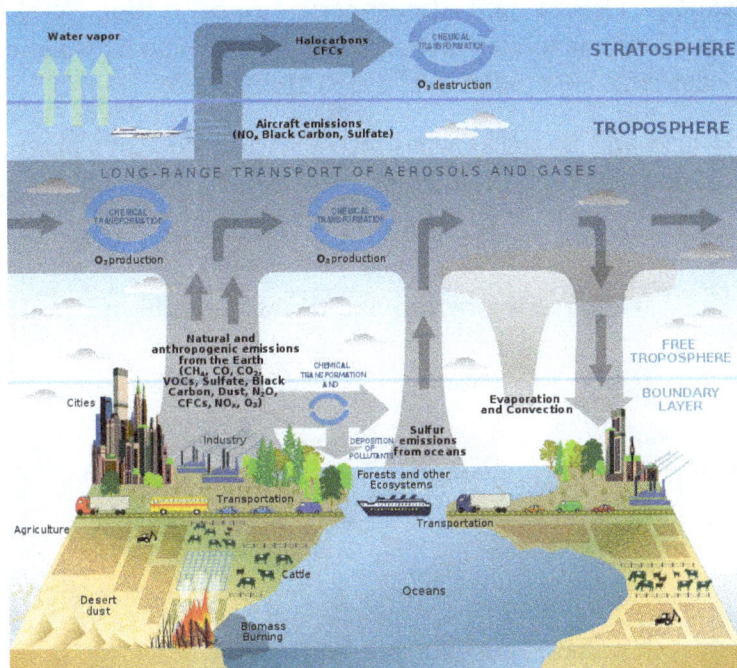

Atmospheric chemistry is a branch of atmospheric science in which the chemistry of the Earth's atmosphere and that of other planets is studied. It is a multidisciplinary field of research and draws on environmental chemistry, physics, meteorology, computer modeling, oceanography, geology and volcanology and other disciplines. Research is increasingly connected with other areas of study such as climatology.

The composition and chemistry of the atmosphere is of importance for several reasons, but primarily because of the interactions between the atmosphere and living organisms. The composition of the Earth's atmosphere has been changed by human activity and some of these changes are harmful to human health, crops and ecosystems. Examples of problems which have been addressed by atmospheric chemistry include acid rain, photochemical smog and global warming. Atmospheric chemistry seeks to understand the causes of these problems, and by obtaining a theoretical understanding of them, allow possible solutions to be tested and the effects of changes in government policy evaluated.

Atmospheric Dynamics

Atmospheric dynamics involves the study of observations and theory dealing with all motion systems of meteorological importance. Common topics studied include diverse phenomena such as thunderstorms, tornadoes, gravity waves, tropical cyclones, extratropical cyclones, jet streams, and global-scale circulations. The goal of dynamical studies is to explain the observed circulations on the basis of fundamental principles from physics. The objectives of such studies incorporate improving weather forecasting, developing methods for predicting seasonal and interannual climate fluctuations, and understanding the implications of human-induced perturbations (e.g., increased carbon dioxide concentrations or depletion of the ozone layer) on the global climate.

In the United Kingdom, atmospheric studies are underpinned by the Meteorological Office. Divisions of the U.S. National Oceanic and Atmospheric Administration (NOAA) oversee research projects and weather modeling involving atmospheric physics. The U.S. National Astronomy and Ionosphere Center also carries out studies of the high atmosphere.

The Earth's magnetic field and the solar wind interact with the atmosphere, creating the ionosphere, Van Allen radiation belts, telluric currents, and radiant energy.

Climatology

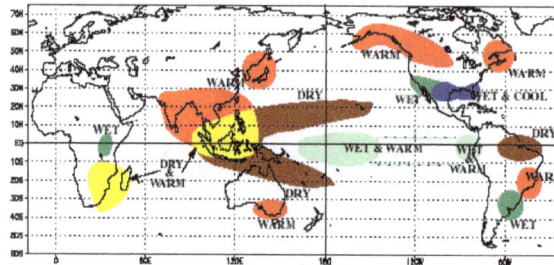

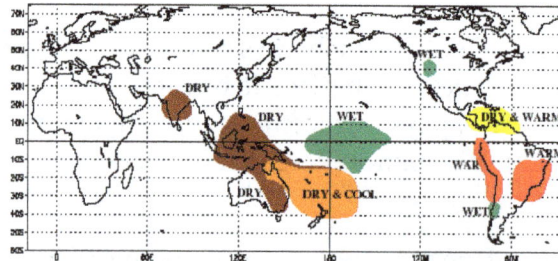

Regional impacts of warm ENSO episodes (El Niño).

In contrast to meteorology, which studies short term weather systems lasting up to a few weeks, climatology studies the frequency and trends of those systems. It studies the periodicity of weather events over years to millennia, as well as changes in long-term average weather patterns, in relation to atmospheric conditions. Climatologists, those who practice climatology, study both the nature of climates – local, regional or global – and the natural or human-induced factors that cause climates to change. Climatology considers the past and can help predict future climate change.

Phenomena of climatological interest include the atmospheric boundary layer, circulation patterns, heat transfer (radiative, convective and latent), interactions between the atmosphere and the oceans and land surface (particularly vegetation, land use and topography), and the chemical and physical composition of the atmosphere. Related disciplines include astrophysics, atmospheric physics, chemistry, ecology, physical geography, geology, geophysics, glaciology, hydrology, oceanography, and volcanology.

Atmospheres on other Celestial Bodies

All of the Solar System's planets have atmospheres. This is because their gravity is strong enough to keep gaseous particles close to the surface. Larger gas giants are massive enough to keep large amounts of the light gases hydrogen and helium close by, while the smaller planets lose these gases into space. The composition of the Earth's atmosphere is different from the other planets because the various life processes that have transpired on the planet have introduced free molecular oxygen. Much of Mercury's atmosphere has been blasted away by the solar wind. The only moon that has retained a dense atmosphere is Titan. There is a thin atmosphere on Triton, and a trace of an atmosphere on the Moon.

Planetary atmospheres are affected by the varying degrees of energy received from either the Sun or their interiors, leading to the formation of dynamic weather systems such as hurricanes, (on Earth), planet-wide dust storms (on Mars), an Earth-sized anticyclone on Jupiter (called the Great Red Spot), and holes in the atmosphere (on Neptune). At least one extrasolar planet, HD 189733 b, has been claimed to possess such a weather system, similar to the Great Red Spot but twice as large.

Hot Jupiters have been shown to be losing their atmospheres into space due to stellar radiation, much like the tails of comets. These planets may have vast differences in temperature between their day and night sides which produce supersonic winds, although the day and night sides of HD 189733b appear to have very similar temperatures, indicating that planet's atmosphere effectively redistributes the star's energy around the planet.

Atmospheric Physics

Atmospheric physics is the application of physics to the study of the atmosphere. Atmospheric physicists attempt to model Earth's atmosphere and the atmospheres of the other planets using fluid flow equations, chemical models, radiation budget, and energy transfer processes in the atmosphere (as well as how these tie into other systems such as the oceans). In order to model weather systems, atmospheric physicists employ elements of scattering theory, wave propagation models, cloud physics, statistical mechanics and spatial statistics which are highly mathematical and related to physics. It has close links to meteorology and climatology and also covers the design and construction of instruments for studying the atmosphere and the interpretation of the data they provide, including remote sensing instruments. At the dawn of the space age and the introduction of sounding rockets, aeronomy became a subdiscipline concerning the upper layers of the atmosphere, where dissociation and ionization are important.

Remote Sensing

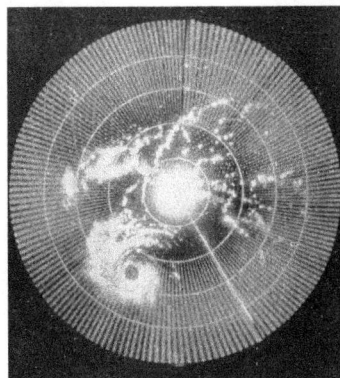

Brightness can indicate reflectivity as in this 1960 weather radar image (of Hurricane Abby). The radar's frequency, pulse form, and antenna largely determine what it can observe.

Remote sensing is the small or large-scale acquisition of information of an object or phenomenon, by the use of either recording or real-time sensing device(s) that is not in physical or intimate contact with the object (such as by way of aircraft, spacecraft, satellite, buoy, or ship). In practice, remote sensing is the stand-off collection through the use of a variety of devices for gathering information on a given object or area which gives more information than sensors at individual sites might convey. Thus, Earth observation or weather satellite collection platforms, ocean and atmospheric observing weather buoy platforms, monitoring of a pregnancy via ultrasound, Magnetic Resonance Imaging (MRI), Positron Emission Tomography (PET), and space probes are all examples of remote sensing. In modern usage, the term generally refers to the use of imaging sensor technologies including but not limited to the use of instruments aboard aircraft and spacecraft, and is distinct from other imaging-related fields such as medical imaging.

There are two kinds of remote sensing. Passive sensors detect natural radiation that is emitted or reflected by the object or surrounding area being observed. Reflected sunlight is the most common source of radiation measured by passive sensors. Examples of passive remote sensors include film photography, infra-red, charge-coupled devices, and radiometers. Active collection, on the other hand, emits energy in order to scan objects and areas whereupon a sensor then detects and measures the radiation that is reflected or backscattered from the target. radar, lidar, and SODAR are examples of active remote sensing techniques used in atmospheric physics where the time delay between emission and return is measured, establishing the location, height, speed and direction of an object.

Remote sensing makes it possible to collect data on dangerous or inaccessible areas. Remote sensing applications include monitoring deforestation in areas such as the Amazon Basin, the effects of climate change on glaciers and Arctic and Antarctic regions, and depth sounding of coastal and ocean depths. Military collection during the cold war made use of stand-off collection of data about dangerous border areas. Remote sensing also replaces costly and slow data collection on the ground, ensuring in the process that areas or objects are not disturbed.

Orbital platforms collect and transmit data from different parts of the electromagnetic spectrum, which in conjunction with larger scale aerial or ground-based sensing and analysis, provides researchers with enough information to monitor trends such as El Niño and other natural long and short term phenomena. Other uses include different areas of the earth sciences such as natural resource management, agricultural fields such as land usage and conservation, and national security and overhead, ground-based and stand-off collection on border areas.

Radiation

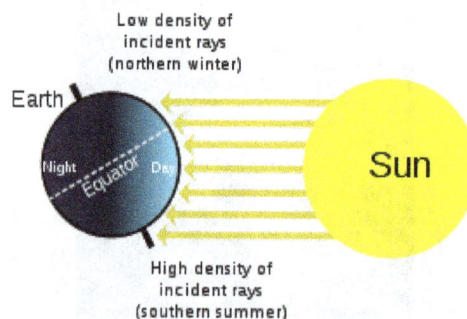

This is a diagram of the seasons. In addition to the density of incident light, the dissipation of light in the atmosphere is greater when it falls at a shallow angle.

Atmospheric physicists typically divide radiation into solar radiation (emitted by the sun) and terrestrial radiation (emitted by Earth's surface and atmosphere).

Solar radiation contains variety of wavelengths. Visible light has wavelengths between 0.4 and 0.7 micrometers. Shorter wavelengths are known as the ultraviolet (UV) part of the spectrum, while longer wavelengths are grouped into the infrared portion of the spectrum. Ozone is most effective in absorbing radiation around 0.25 micrometers, where UV-c rays lie in the spectrum. This increases the temperature of the nearby stratosphere. Snow reflects 88% of UV rays, while sand reflects 12%, and water reflects only 4% of incoming UV radiation. The more glancing the angle is between the atmosphere and the sun's rays, the more likely that energy will be reflected or absorbed by the atmosphere.

Terrestrial radiation is emitted at much longer wavelengths than solar radiation. This is because Earth is much colder than the sun. Radiation is emitted by Earth across a range of wavelengths, as formalized in Planck's law. The wavelength of maximum energy is around 10 micrometers.

Cloud Physics

Cloud physics is the study of the physical processes that lead to the formation, growth and precipitation of clouds. Clouds are composed of microscopic droplets of water (warm clouds), tiny crystals of ice, or both (mixed phase clouds). Under suitable conditions, the droplets combine to form precipitation, where they may fall to the earth. The precise mechanics of how a cloud forms and grows is not completely understood, but scientists have developed theories explaining the structure of clouds by studying the microphysics of individual droplets. Advances in radar and satellite technology have also allowed the precise study of clouds on a large scale.

Atmospheric Electricity

Cloud to ground Lightning in the global atmospheric electrical circuit.

Atmospheric electricity is the term given to the electrostatics and electrodynamics of the atmosphere (or, more broadly, the atmosphere of any planet). The Earth's surface, the ionosphere, and the atmosphere is known as the global atmospheric electrical circuit. Lightning discharges 30,000 amperes, at up to 100 million volts, and emits light, radio waves, x-rays and even gamma rays. Plasma temperatures in lightning can approach 28,000 kelvins and electron densities may exceed $10^{24}/m^3$.

Atmospheric Tide

The largest-amplitude atmospheric tides are mostly generated in the troposphere and stratosphere when the atmosphere is periodically heated as water vapour and ozone absorb solar radiation during the day. The tides generated are then able to propagate away from these source regions and ascend into the mesosphere and thermosphere. Atmospheric tides can be measured as regular fluctuations in wind, temperature, density and pressure. Although atmospheric tides share much in common with ocean tides they have two key distinguishing features:

i) Atmospheric tides are primarily excited by the Sun's heating of the atmosphere whereas ocean tides are primarily excited by the Moon's gravitational field. This means that most atmospheric tides have periods of oscillation related to the 24-hour length of the solar day whereas ocean tides have longer periods of oscillation related to the lunar day (time between successive lunar transits) of about 24 hours 51 minutes.

ii) Atmospheric tides propagate in an atmosphere where density varies significantly with height. A consequence of this is that their amplitudes naturally increase exponentially as the tide ascends into progressively more rarefied regions of the atmosphere. In contrast, the density of the oceans varies only slightly with depth and so there the tides do not necessarily vary in amplitude with depth.

Note that although solar heating is responsible for the largest-amplitude atmospheric tides, the gravitational fields of the Sun and Moon also raise tides in the atmosphere, with the lunar gravitational atmospheric tidal effect being significantly greater than its solar counterpart.

At ground level, atmospheric tides can be detected as regular but small oscillations in surface pressure with periods of 24 and 12 hours. Daily pressure maxima occur at 10 a.m. and 10 p.m. local time, while minima occur at 4 a.m. and 4 p.m. local time. The absolute maximum occurs at 10 a.m. while the absolute minimum occurs at 4 p.m. However, at greater heights the amplitudes of the tides can become very large. In the mesosphere (heights of ~ 50 – 100 km) atmospheric tides can reach amplitudes of more than 50 m/s and are often the most significant part of the motion of the atmosphere.

Aeronomy

Aeronomy is the science of the upper region of the atmosphere, where dissociation and ionization are important. The term aeronomy was introduced by Sydney Chapman in 1960. Today, the term also includes the science of the corresponding regions of the atmospheres of other planets. Research in aeronomy requires access to balloons, satellites, and sounding rockets which provide valuable data about this region of the atmosphere. Atmospheric tides play an important role in interacting with both the lower and upper atmosphere. Amongst the phenomena studied are upper-atmospheric lightning discharges, such as luminous events called red sprites, sprite halos, blue jets, and elves.

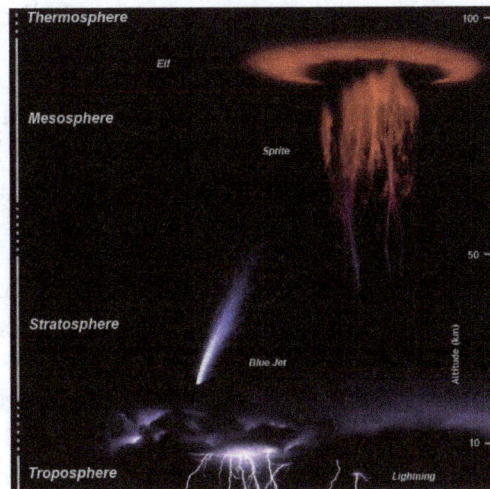

Representation of upper-atmospheric lightning and electrical-discharge phenomena.

Centers of Research

In the UK, atmospheric studies are underpinned by the Met Office, the Natural Environment Research Council and the Science and Technology Facilities Council. Divisions of the U.S. National Oceanic and Atmospheric Administration (NOAA) oversee research projects and weather modeling involving atmospheric physics. The US National Astronomy and Ionosphere Center also carries out studies of the high atmosphere. In Belgium, the Belgian Institute for Space Aeronomy studies the atmosphere and outer space.

Atmospheric Electricity

Atmospheric electricity is the study of electrical charges in the Earth's atmosphere (or that of another planet). The movement of charge between the Earth's surface, the atmosphere, and the ionosphere is known as the global atmospheric electrical circuit. Atmospheric electricity is an interdisciplinary topic, involving concepts from electrostatics, atmospheric physics, meteorology and Earth science.

Thunderstorms act as a giant battery in the atmosphere, charging up the ionosphere to about 400,000 volts with respect to the surface. This sets up an electric field throughout the atmosphere, which decreases with increase in altitude. Atmospheric ions created by cosmic rays and natural radioactivity move in the electric field, so a very small current flows through the atmosphere, even away from thunderstorms. Near the surface of the earth, the magnitude of the field is around 100 V/m.

Atmospheric electricity involves both thunderstorms, which create lightning bolts to rapidly discharge huge amounts of atmospheric charge stored in storm clouds, and the continual electrification of the air due to ionization from cosmic rays and natural radioactivity, which ensure that the atmosphere is never quite neutral.

History

Sparks drawn from electrical machines and from Leyden jars suggested to the early experimenters, Hauksbee, Newton, Wall, Nollet, and Gray, that lightning was caused by electric discharges. In

1708, Dr. William Wall was one of the first to observe that spark discharges resembled miniature lightning, after observing the sparks from a charged piece of amber.

Benjamin Franklin's experiments showed that electrical phenomena of the atmosphere were not fundamentally different from those produced in the laboratory, by listing many similarities between electricity and lightning. By 1749, Franklin observed lightning to possess almost all the properties observable in electrical machines.

In July 1750, Franklin hypothesized that electricity could be taken from clouds via a tall metal aerial with a sharp point. Before Franklin could carry out his experiment, in 1752 Thomas-François Dalibard erected a 40-foot (12 m) iron rod at Marly-la-Ville, near Paris, drawing sparks from a passing cloud. With ground-insulated aerials, an experimenter could bring a grounded lead with an insulated wax handle close to the aerial, and observe a spark discharge from the aerial to the grounding wire. In May 1752, Dalibard affirmed that Franklin's theory was correct.

Around June 1752, Franklin reportedly performed his famous kite experiment. The kite experiment was repeated by Romas, who drew from a metallic string sparks 9 feet (2.7 m) long, and by Cavallo, who made many important observations on atmospheric electricity. Lemonnier (1752) also reproduced Franklin's experiment with an aerial, but substituted the ground wire with some dust particles (testing attraction). He went on to document the *fair weather condition*, the clear-day electrification of the atmosphere, and its diurnal variation. Beccaria (1775) confirmed Lemonnier's diurnal variation data and determined that the atmosphere's charge polarity was positive in fair weather. Saussure (1779) recorded data relating to a conductor's induced charge in the atmosphere. Saussure's instrument (which contained two small spheres suspended in parallel with two thin wires) was a precursor to the electrometer. Saussure found that the atmospheric electrification under clear weather conditions had an annual variation, and that it also varied with height. In 1785, Coulomb discovered the electrical conductivity of air. His discovery was contrary to the prevailing thought at the time, that the atmospheric gases were insulators (which they are to some extent, or at least not very good conductors when not ionized). Erman (1804) theorized that the Earth was negatively charged, and Peltier (1842) tested and confirmed Erman's idea.

Several researchers contributed to the growing body of knowledge about atmospheric electrical phenomena. Francis Ronalds began observing the potential gradient and air-earth currents around 1810, including making continuous automated recordings. He resumed his research in the 1840s as the inaugural Honorary Director of the Kew Observatory, where the first extended and comprehensive dataset of electrical and associated meteorological parameters was created. He also supplied his equipment to other facilities around the world with the goal of delineating atmospheric electricity on a global scale. Kelvin's new water dropper collector and divided-ring electrometer were introduced at Kew Observatory in the 1860s, and atmospheric electricity remained a speciality of the observatory until its closure. For high-altitude measurements, kites were once used, and weather balloons or aerostats are still used, to lift experimental equipment into the air. Early experimenters even went aloft themselves in hot-air balloons.

Hoffert (1888) identified individual lightning downward strokes using early cameras. Elster and Geitel, who also worked on thermionic emission, proposed a theory to explain thunderstorms' electrical structure (1885) and, later, discovered atmospheric radioactivity (1899) from the existence of positive and negative ions in the atmosphere. Pockels (1897) estimated lightning current intensity by analyzing lightning flashes in basalt (c. 1900) and studying the left-over magnetic fields caused by lightning. Discoveries about the electrification of the atmosphere via sensitive electrical instruments and ideas on how the Earth's negative charge is maintained were developed mainly in the 20th century, with CTR Wilson playing an important part. Current research on atmospheric electricity focuses mainly on lightning, particularly high-energy particles and transient luminous events, and the role of non-thunderstorm electrical processes in weather and climate.

Nikola Tesla and Hermann Plauson investigated the production of energy and power via atmospheric electricity. Tesla also proposed to use the atmospheric electrical circuit to transceive wireless energy over large distances, but no feasible apparatus to extract energy from atmospheric electricity has been built.

Description

Atmospheric electricity is always present, and during fine weather away from thunderstorms, the air above the surface of Earth is positively charged, while the Earth's surface charge is negative. It can be understood in terms of a difference of potential between a point of the Earth's surface, and a point somewhere in the air above it. Because the atmospheric electric field is negatively directed in fair weather, the convention is to refer to the potential gradient, which has the opposite sign and is about 100V/m at the surface. There is a weak conduction current of atmospheric ions moving in the atmospheric electric field, about 2 picoAmperes per square metre, and the air is weakly conductive due to the presence of these atmospheric ions.

Variations

Global daily cycles in the atmospheric electric field, with a minimum around 03 UT and peaking roughly 16 hours later, were researched by the Carnegie Institution of Washington in the 20th century. This Carnegie curve variation has been described as "the fundamental electrical heartbeat of the planet".

Even away from thunderstorms, atmospheric electricity can be highly variable, but, generally, the electric field is enhanced in fogs and dust whereas the atmospheric electrical conductivity is diminished.

Near Space

The electrosphere layer (from tens of kilometers above the surface of the earth to the ionosphere) has a high electrical conductivity and is essentially at a constant electric potential. The ionosphere is the inner edge of the magnetosphere and is the part of the atmosphere that is ionized by solar radiation. (Photoionization is a physical process in which a photon is incident on an atom, ion or molecule, resulting in the ejection of one or more electrons.)

Cosmic Radiation

The Earth, and almost all living things on it, are constantly bombarded by radiation from outer space. This radiation primarily consists of positively charged ions from protons to iron and larger nuclei derived sources outside our solar system. This radiation interacts with atoms in the atmosphere to create an air shower of secondary ionising radiation, including X-rays, muons, protons, alpha particles, pions, and electrons. Ionization from this secondary radiation ensures that the atmosphere is weakly conductive, and that the slight current flow from these ions over the Earth's surface balances the current flow from thunderstorms. Ions have characteristic parameters such as mobility, lifetime, and generation rate that vary with altitude.

Earth-Ionosphere Cavity

The Potential difference between the ionosphere and the Earth is maintained by thunderstorms. In the Earth-ionosphere cavity, the electric field and conduction current in the lower atmosphere are primarily controlled by ions.

The Schumann resonance is a set of spectrum peaks in the extremely low frequency (ELF) portion of the Earth's electromagnetic field spectrum. Schumann resonance is due to the space between the surface of the Earth and the conductive ionosphere acting as a waveguide. The limited dimensions of the earth cause this waveguide to act as a resonant cavity for electromagnetic waves. The cavity is naturally excited by energy from lightning strikes.

Thunderstorms and Lightning

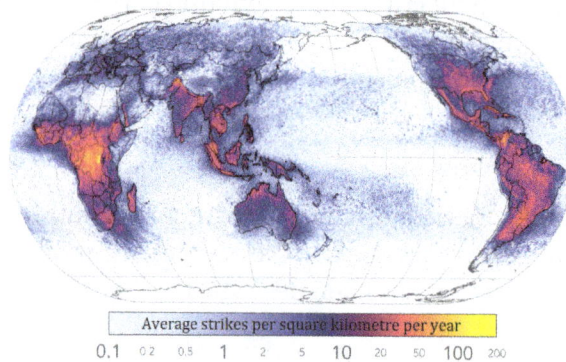

World map showing frequency of lightning strikes, in flashes per km² per year (equal-area projection). Lightning strikes most frequently in the Democratic Republic of the Congo. Combined 1995–2003 data from the Optical Transient Detector and 1998–2003 data from the Lightning Imaging Sensor.

If the quantity of water that is condensed in and subsequently precipitated from a cloud is known, then the total energy of a thunderstorm can be calculated. In an average thunderstorm, the energy released amounts to about 10,000,000 kilowatt-hours (3.6×10^{13} joule), which is equivalent to a 20-kiloton nuclear warhead. A large, severe thunderstorm might be 10 to 100 times more energetic.

Collisions between ice and soft hail (graupel) inside cumulonimbus clouds causes separation of positive and negative charges within the cloud, essential for the generation of lightning. How lightning initially forms is still a matter of debate: Scientists have studied root causes ranging from atmospheric perturbations (wind, humidity, and atmospheric pressure) to the impact of solar wind and energetic particles.

An average bolt of lightning carries a negative electric current of 40 kiloamperes (kA) (although some bolts can be up to 120 kA), and transfers a charge of five coulombs and energy of 500 MJ, or enough energy to power a 100-watt lightbulb for just under two months. The voltage depends on the length of the bolt, with the dielectric breakdown of air being three million volts per meter, and lightning bolts often being several hundred meters long. However, lightning leader development is not a simple matter of dielectric breakdown, and the ambient electric fields required for lightning leader propagation can be a few orders of magnitude less than dielectric breakdown strength. Further, the potential gradient inside a well-developed return-stroke channel is on the order of hundreds of volts per meter or less due to intense channel ionization, resulting in a true power output on the order of megawatts per meter for a vigorous return-stroke current of 100 kA .

Lightning sequence (Duration: 0.32 seconds)

Corona Discharges

St. Elmo's Fire is an electrical phenomenon in which luminous plasma is created by a coronal discharge originating from a grounded object. Ball lightning is often erroneously identified as St. Elmo's Fire, whereas they are separate and distinct phenomena. Although referred to as "fire", St. Elmo's Fire is, in fact, plasma, and is observed, usually during a thunderstorm, at the tops of trees, spires or other tall objects, or on the heads of animals, as a brush or star of light.

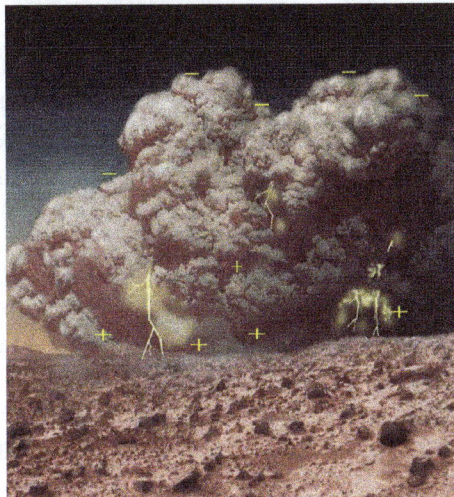

A depiction of atmospheric electricity in a Martian dust storm, which has been suggested as a possible explanation for enigmatic chemistry results from Mars.

Corona is caused by the electric field around the object in question ionizing the air molecules, producing a faint glow easily visible in low-light conditions. Approximately 1,000 – 30,000 volts per centimetre is required to induce St. Elmo's Fire; however, this is dependent on the geometry of the object in question. Sharp points tend to require lower voltage levels to produce the same result because electric fields are more concentrated in areas of high curvature, thus discharges are more intense at the end of pointed objects. St. Elmo's Fire and normal sparks both can appear when high electrical voltage affects a gas. St. Elmo's fire is seen during thunderstorms when the ground below the storm is electrically charged, and there is high voltage in the air between the cloud and the ground. The voltage tears apart the air molecules and the gas begins to glow. The nitrogen and oxygen in the Earth's atmosphere causes St. Elmo's Fire to fluoresce with blue or violet light; this is similar to the mechanism that causes neon signs to glow.

Electrical system grounding

Atmospheric charges can cause undesirable, dangerous, and potentially lethal charge potential buildup in suspended electric wire power distribution systems. Bare wires suspended in the air spanning many kilometers and isolated from the ground can collect very large stored capacitance at high voltage static charge potentials, even when there is no thunderstorm or lightning occurring. This charge potential will seek to discharge itself through the path of least insulation, which can occur when a person reaches out to activate a power switch or to use an electric device.

To dissipate atmospheric charge buildup, one side of the electrical distribution system is connected to the earth at many points throughout the distribution system, as often as on every support pole. The one earth-connected wire is commonly referred to as the "protective earth", and provides path for the charge potential to dissipate without causing damage, and provides redundancy in case any one of the ground paths is poor due to corrosion or poor ground conductivity. The additional electric grounding wire that carries no power serves a secondary role, providing a high-current short-circuit path to rapidly blow fuses and render a damaged device safe, rather than have an ungrounded device with damaged insulation become "electrically live" via the grid power supply, and hazardous to touch.

Each transformer in an alternating current distribution grid segments the grounding system into a new separate circuit loop. These separate grids must also be grounded on one side to prevent charge buildup within them relative to the rest of the system, and which could cause damage from charge potentials discharging across the transformer coils to the other grounded side of the distribution network.

Atmospheric Tide

Atmospheric tides are global-scale periodic oscillations of the atmosphere. In many ways they are analogous to ocean tides. Atmospheric tides can be excited by:

- The regular day–night cycle in the Sun's heating of the atmosphere (insolation)

- The gravitational field pull of the Moon

- Non-linear interactions between tides and planetary waves.

- Large-scale latent heat release due to deep convection in the tropics.

General Characteristics

The largest-amplitude atmospheric tides are mostly generated in the troposphere and stratosphere when the atmosphere is periodically heated, as water vapor and ozone absorb solar radiation during the day. These tides propagate away from the source regions and ascend into the mesosphere and thermosphere. Atmospheric tides can be measured as regular fluctuations in wind, temperature, density and pressure. Although atmospheric tides share much in common with ocean tides they have two key distinguishing features:

1. Atmospheric tides are primarily excited by the Sun's heating of the atmosphere whereas ocean tides are excited by the Moon's gravitational pull and to a lesser extent by the Sun's gravity. This means that most atmospheric tides have periods of oscillation related to the 24-hour length of the solar day whereas ocean tides have periods of oscillation related both to the solar day as well as to the longer lunar day (time between successive lunar transits) of about 24 hours 51 minutes.

2. Atmospheric tides propagate in an atmosphere where density varies significantly with height. A consequence of this is that their amplitudes naturally increase exponentially as the tide ascends into progressively more rarefied regions of the atmosphere. In contrast, the density of the oceans varies only slightly with depth and so there the tides do not necessarily vary in amplitude with depth.

At ground level, atmospheric tides can be detected as regular but small oscillations in surface pressure with periods of 24 and 12 hours. However, at greater heights, the amplitudes of the tides can become very large. In the mesosphere (heights of ~ 50–100 km) atmospheric tides can reach amplitudes of more than 50 m/s and are often the most significant part of the motion of the atmosphere.

The reason for this dramatic growth in amplitude from tiny fluctuations near the ground to oscillations that dominate the motion of the mesosphere lies in the fact that the density of the atmosphere decreases with increasing height. As tides or waves propagate upwards, they move into regions of lower and lower density. If the tide or wave is not dissipating, then its kinetic energy density must be conserved. Since the density is decreasing, the amplitude of the tide or wave increases correspondingly so that energy is conserved.

Following this growth with height atmospheric tides have much larger amplitudes in the middle and upper atmosphere than they do at ground level.

Solar Atmospheric Tides

The largest amplitude atmospheric tides are generated by the periodic heating of the atmosphere by the Sun – the atmosphere is heated during the day and not heated at night. This regular diurnal (daily) cycle in heating generates tides that have periods related to the solar day. It might initially be expected that this diurnal heating would give rise to tides with a period of 24 hours, corresponding to the heating's periodicity. However, observations reveal that large amplitude tides are generated with periods of 24 and 12 hours. Tides have also been observed with periods of 8 and 6 hours, although these latter tides generally have smaller amplitudes. This set of periods occurs because the solar heating of the atmosphere occurs in an approximate square wave profile and so is rich in harmonics. When this pattern is decomposed into separate frequency components using a Fourier transform, as well as the mean and daily (24-hr) variation, significant oscillations with periods of 12, 8 and 6 hrs are produced. Tides generated by the gravitational effect of the sun are very much smaller than those generated by solar heating. Solar tides will refer to only thermal solar tides from this point.

Solar energy is absorbed throughout the atmosphere some of the most significant in this context are water vapor at (~0–15 km) in the troposphere, ozone at (~30 to 60 km) in the stratosphere and molecular oxygen and molecular nitrogen at (~120 to 170 km) in the thermosphere. Variations in the global distribution and density of these species result in changes in the amplitude of the solar tides. The tides are also affected by the environment through which they travel.

Solar tides can be separated into two components: migrating and non-migrating.

Migrating Solar Tides

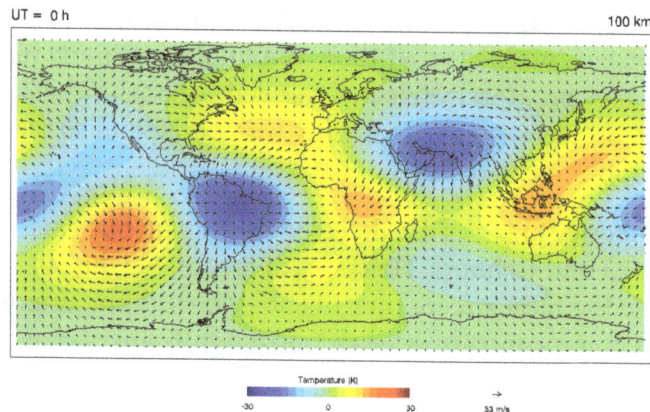

Tidal temperature and wind perturbations at 100 km altitude for September 2005 as a function of universal time. The image is based upon observations from the SABER and TIDI instruments on board the TIMED satellite. It shows the superposition of the most important diurnal and semidiurnal tidal components
(migrating + nonmigrating).

Migrating tides are sun synchronous – from the point of view of a stationary observer on the ground they propagate westwards with the apparent motion of the sun. As the migrating tides stay fixed relative to the sun a pattern of excitation is formed that is also fixed relative to the Sun. Changes in the tide observed from a stationary viewpoint on the Earth's surface are caused by the rotation of the Earth with respect to this fixed pattern. Seasonal variations of the tides also occur as the Earth tilts relative to the Sun and so relative to the pattern of excitation.

The migrating solar tides have been extensively studied both through observations and mechanistic models.

Non-migrating Solar Tides

Non-migrating tides can be thought of as global-scale waves with the same periods as the migrating tides. However, non-migrating tides do not follow the apparent motion of the sun. Either they do not propagate horizontally, they propagate eastwards or they propagate westwards at a different speed to the sun. These non-migrating tides may be generated by differences in topography with longitude, land-sea contrast, and surface interactions. An important source is latent heat release due to deep convection in the tropics.

The primary source for the 24-hr tide is in the lower atmosphere where surface effects are important. This is reflected in a relatively large non-migrating component seen in longitudinal differences in tidal amplitudes. Largest amplitudes have been observed over South America, Africa and Australia.

Lunar Atmospheric Tides

Atmospheric tides are also produced through the gravitational effects of the Moon. *Lunar (gravitational) tides* are much weaker than *solar (thermal) tides* and are generated by the motion of the Earth's oceans (caused by the Moon) and to a lesser extent the effect of the Moon's gravitational attraction on the atmosphere.

Classical Tidal Theory

The basic characteristics of the atmospheric tides are described by the *classical tidal theory*. By neglecting mechanical forcing and dissipation, the classical tidal theory assumes that atmospheric wave motions can be considered as linear perturbations of an initially motionless zonal mean state that is horizontally stratified and isothermal. The two major results of the classical theory are:

- atmospheric tides are eigenmodes of the atmosphere described by Hough functions
- amplitudes grow exponentially with height.

Basic Equations

The primitive equations lead to the linearized equations for perturbations (primed variables) in a spherical isothermal atmosphere:

- horizontal momentum equations

$$\frac{\partial u'}{\partial t} - 2\Omega \sin\varphi\, v' + \frac{1}{a\cos\varphi}\frac{\partial \Phi'}{\partial \lambda} = 0$$

$$\frac{\partial v'}{\partial t} + 2\Omega \sin\varphi\, u' + \frac{1}{a}\frac{\partial \Phi'}{\partial \varphi} = 0$$

- energy equation

$$\frac{\partial^2}{\partial t \partial z}\Phi' + N^2 w' = \frac{\kappa J'}{H}$$

- continuity equation

$$\frac{1}{a\cos\varphi}\left(\frac{\partial u'}{\partial \lambda} + \frac{\partial}{\partial \varphi}(v'\cos\varphi)\right) + \frac{1}{\varrho_o}\frac{\partial}{\partial z}(\varrho_o w') = 0$$

with the definitions

- u eastward zonal wind
- v northward meridional wind
- w upward vertical wind
- Φ geopotential, $\int g(z,\varphi)dz$

- N^2 square of Brunt-Vaisala (buoyancy) frequency

- Ω angular velocity of the Earth

- ϱ_o density $\propto \exp(-z/H)$

- z altitude

- λ geographic longitude

- φ geographic latitude

- J heating rate per unit mass

- a radius of the Earth

- g gravity acceleration

- H constant scale height

- t time

Separation of Variables

The set of equations can be solved for *atmospheric tides*, i.e., longitudinally propagating waves of zonal wavenumber s and frequency σ. Zonal wavenumber s is a positive integer so that positive values for σ correspond to eastward propagating tides and negative values to westward propagating tides. A separation approach of the form

$$\Phi'(\varphi,\lambda,z,t) = \hat{\Phi}(\varphi,z)e^{i(s\lambda-\sigma t)}$$

$$\hat{\Phi}(\varphi,z) = \sum_n \Theta_n(\varphi)G_n(z)$$

and doing some math yields expressions for the latitudinal and vertical structure of the tides.

Laplace's Tidal Equation

The latitudinal structure of the tides is described by the *horizontal structure equation* which is also called *Laplace's tidal equation*:

$$L\Theta_n + \varepsilon_n \Theta_n = 0$$

with *Laplace operator*

$$L = \frac{\partial}{\partial\mu}\left[\frac{(1-\mu^2)}{(\eta^2-\mu^2)}\frac{\partial}{\partial\mu}\right] - \frac{1}{\eta^2-\mu^2}\left[-\frac{s(\eta^2+\mu^2)}{\eta(\eta^2-\mu^2)}+\frac{s^2}{1-\mu^2}\right]$$

using $\mu = \sin\varphi$, $\eta = \sigma/(2\Omega)$ and *eigenvalue*

$$\varepsilon_n = (2\Omega a)^2 / gh_n.$$

Hence, atmospheric tides are eigenoscillations (eigenmodes) of Earth's atmosphere with eigenfunctions Θ_n, called Hough functions, and eigenvalues ε_n. The latter define the *equivalent depth* h_n which couples the latitudinal structure of the tides with their vertical structure.

General Solution of Laplaces Equation

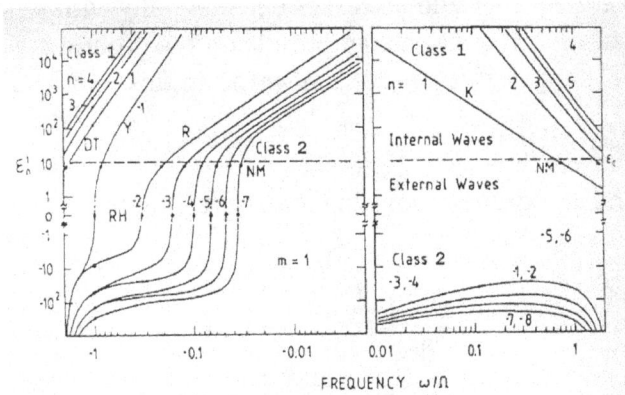

Eigenvalue ε of wave modes of zonal wave number s = 1 vs. normalized frequency $\nu = \omega/\Omega$ where $\Omega = 7.27 \times 10^{-5}$ s^{-1} is the angular frequency of one solar day. Waves with positive (negative) frequencies propagate to the east (west). The horizontal dashed line is at $\varepsilon_c \simeq 11$ and indicates the transition from internal to external waves . Meaning of the symbols: 'RH' Rossby-Haurwitz waves (ε = 0); 'Y' Yanai waves; 'K' Kelvin waves; 'R' Rossby waves; 'DT' Diurnal tides (ν = -1); 'NM' Normal modes ($\varepsilon \simeq \varepsilon_c$).

Longuet-Higgins has completely solved Laplace's equations and has discovered tidal modes with negative eigenvalues ε_n^s (Figure above). There exist two kinds of waves: class 1 waves, (sometimes called gravity waves), labelled by positive n, and class 2 waves (sometimes called rotational waves), labelled by negative n. Class 2 waves owe their existence to the Coriolis force and can only exist for periods greater than 12 hours (or $|\nu| \leq 2$). Tidal waves can be either internal (travelling waves) with positive eigenvalues (or equivalent depth) which have finite vertical wavelengths and can transport wave energy upward, or external (evanescent waves) with negative eigenvalues and infinitely large vertical wavelengths meaning that their phases remain constant with altitude. These external wave modes cannot transport wave energy, and their amplitudes decrease exponentially with height outside their source regions. Even numbers of n correspond to waves symmetric with respect to the equator, and odd numbers corresponding to antisymmetric waves. The transition from internal to external waves appears at $\varepsilon \simeq \varepsilon_c$, or at the vertical wavenumber k$_z$ = 0, and $\lambda_z \Rightarrow \infty$, respectively.

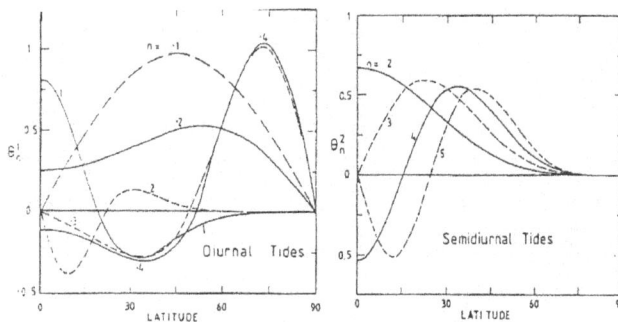

Pressure amplitudes vs. latitude of the Hough functions of the diurnal tide (s = 1; ν = -1) (left) and of the semidiural tides (s = 2; ν = -2) (right) on the northern hemisphere. Solid curves: symmetric waves; dashed curves: antisymmetric waves.

The fundamental solar diurnal tidal mode which optimally matches the solar heat input configuration and thus is most strongly excited is the Hough mode (1, -2) (Figure above). It depends on local time and travels westward with the Sun. It is an external mode of class 2 and has the eigenvalue of $\varepsilon_{-2}^{1} = -12.56$. Its maximum pressure amplitude on the ground is about 60 hPa. The largest solar semidiurnal wave is mode (2, 2) with maximum pressure amplitudes at the ground of 120 hPa. It is an internal class 1 wave. Its amplitude increases exponentially with altitude. Although its solar excitation is half of that of mode (1, −2), its amplitude on the ground is larger by a factor of two. This indicates the effect of suppression of external waves, in this case by a factor of four.

Vertical Structure Equation

For bounded solutions and at altitudes above the forcing region, the *vertical structure equation* in its canonical form is:

$$\frac{\partial^2 G_n^{\star}}{\partial x^2} + \alpha_n^2 G_n^{\star} = F_n(x)$$

with solution

$$G_n^{\star}(x) \sim \begin{cases} e^{-|\alpha_n|x} & : \alpha_n^2 < 0, \text{ evanescent or trapped} \\ e^{i\alpha_n x} & : \alpha_n^2 > 0, \text{ propagating} \\ e^{\left(\kappa - \frac{1}{2}\right)x} & : h_n = H/(1-\kappa), F_n(x) = 0 \forall x, \text{ Lamb waves (free solutions)} \end{cases}$$

using the definitions

$$\alpha_n^2 = \kappa H / h_n - 1/4$$

$$x = z/H$$

$$G_n^{\star} \quad G_n \varrho_o^{1/2} N^{-1}$$

$$F_n(x) = -\frac{\varrho_o^{-1/2}}{i\sigma N} \frac{\partial}{\partial x}(\varrho_o J_n).$$

Propagating Solutions

Therefore, each wavenumber/frequency pair (a tidal *component*) is a superposition of associated Hough functions (often called tidal *modes* in the literature) of index n. The nomenclature is such that a negative value of n refers to evanescent modes (no vertical propagation) and a positive value to propagating modes. The equivalent depth h_n is linked to the vertical wavelength $\lambda_{z,n}$, since α_n / H is the vertical wavenumber:

$$\lambda_{z,n} = \frac{2\pi H}{\alpha_n} = \frac{2\pi H}{\sqrt{\dfrac{\kappa H}{h_n} - \dfrac{1}{4}}}.$$

For propagating solutions $(\alpha_n^2 > 0)$, the vertical group velocity

$$c_{gz,n} = H \frac{\partial \sigma}{\partial \alpha_n}$$

becomes positive (upward energy propagation) only if $\alpha_n > 0$ for westward $(\sigma < 0)$ or if $\alpha_n < 0$ for eastward $(\sigma > 0)$ propagating waves. At a given height $x = z / H$, the wave maximizes for

$$K_n = s\lambda + \alpha_n x - \sigma t = 0.$$

For a fixed longitude λ, this in turn always results in downward phase progression as time progresses, independent of the propagation direction. This is an important result for the interpretation of observations: *downward phase progression in time means an upward propagation of energy and therefore a tidal forcing lower in the atmosphere.* Amplitude increases with height $\propto e^{z/2H}$, as density decreases.

Dissipation

Damping of the tides occurs primarily in the lower thermosphere region, and may be caused by turbulence from breaking gravity waves. A similar phenomena to ocean waves breaking on a beach, the energy dissipates into the background atmosphere. Molecular diffusion also becomes increasingly important at higher levels in the lower thermosphere as the mean free path increases in the rarefied atmosphere.

At thermospheric heights, attenuation of atmospheric waves, mainly due to collisions between the neutral gas and the ionospheric plasma, becomes significant so that at above about 150 km altitude, all wave modes gradually become external waves, and the Hough functions degenerate to spherical functions; e.g., mode (1, -2) develops to the spherical function $P_1^1(\theta)$, mode (2, 2) becomes $P_2^2(\theta)$, with θ the co-latitude, etc. . Within the thermosphere, mode (1, -2) is the predominant mode reaching diurnal temperature amplitudes at the exosphere of at least 140 K and horizontal winds of the order of 100 m/s and more increasing with geomagnetic activity. It is responsible for the electric Sq currents within the ionospheric dynamo region between about 100 and 200 km altitude.

Effects of Atmospheric Tide

The tides form an important mechanism for transporting energy from the lower atmosphere into the upper atmosphere, while dominating the dynamics of the mesosphere and lower thermosphere. Therefore, understanding the atmospheric tides is essential in understanding the atmosphere as a whole. Modeling and observations of atmospheric tides are needed in order to monitor and predict changes in the Earth's atmosphere.

Aeronomy

Aeronomy is the meteorological science of the upper region of the Earth's or other planetary atmospheres, which relates to the atmospheric motions, its chemical composition and properties, and the reaction to it from the environment from space. The term *aeronomy* was introduced by Sydney Chapman in a Letter to the Editor of *Nature* entitled *Some Thoughts on Nomenclature* in 1946. Studies within the subject also investigates the causes of dissociation or ionization processes.

Today the term also includes the science of the corresponding regions of the atmospheres of other planets. Aeronomy is a branch of atmospheric physics. Research in aeronomy requires access to balloons, satellites, and sounding rockets which provide valuable data about this region of the atmosphere. Atmospheric tides dominate the dynamics of the mesosphere and lower thermosphere, essential to understanding the atmosphere as a whole. Other phenomena studied are upper-atmospheric lightning discharges, such as red sprites, sprite halos or blue jets.

Atmospheric Tides

Atmospheric tides are global-scale periodic oscillations of the atmosphere. In many ways they are analogous to ocean tides. Atmospheric tides form an important mechanism for transporting energy input into the lower atmosphere from the upper atmosphere, while dominating the dynamics of the mesosphere and lower thermosphere. Therefore, learning about atmospheric tides is essential in understanding the atmosphere as a whole. Modeling and observations of atmospheric tides are needed in order to monitor and predict changes in the Earth's atmosphere.

Upper-atmospheric Lightning

Upper-atmospheric lightning or upper-atmospheric discharge are terms sometimes used by researchers to refer to a family of electrical-breakdown phenomena that occur well above the altitudes of normal lightning. The preferred current usage is transient luminous events (TLEs) to refer to the various types of electrical-discharge phenomena induced in the upper atmosphere by tropospheric lightning. TLEs includes red sprites, sprite halos, blue jets, and elves.

Diurnal Temperature Variation

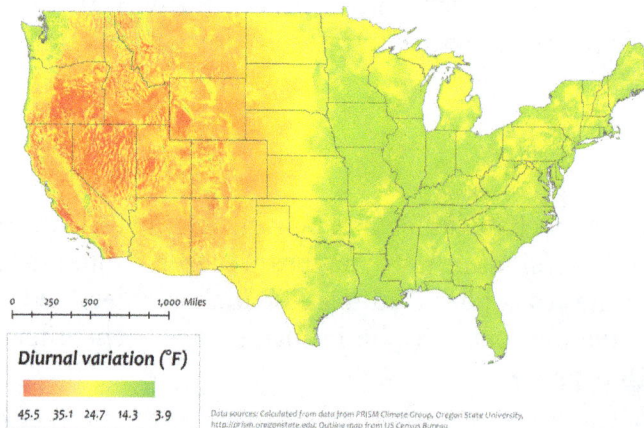

Average Diurnal Variation of Temperature in July

Map of diurnal temperature variation for the month of July in the contiguous United States

In meteorology, diurnal temperature variation is the variation between a high temperature and a low temperature that occurs during the same day.

Temperature Lag

Temperature lag is an important factor in diurnal temperature variation: peak daily temperature generally occurs *after* noon, as air keeps net absorbing heat even after noon, and similarly minimum daily temperature generally occurs substantially after midnight, indeed occurring during early morning in the hour around dawn, since heat is lost all night long. The analogous annual phenomenon is seasonal lag.

As solar energy strikes the earth's surface each morning, a shallow 1–3-centimetre (0.39–1.18 in) layer of air directly above the ground is heated by conduction. Heat exchange between this shallow layer of warm air and the cooler air above is very inefficient. On a warm summer's day, for example, air temperatures may vary by 16.5°C (30°F) from just above the ground to waist height. Incoming solar radiation exceeds outgoing heat energy for many hours after noon and equilibrium is usually reached from 3–5 p.m. but this may be affected by a variety of different things such as large bodies of water, soil type and cover, wind, cloud cover/water vapor, and moisture on the ground.

Differences in Variation

Diurnal temperature variations are greatest very near the earth's surface.

High desert areas typically have the greatest diurnal temperature variations. Low lying, humid areas typically have the least. This explains why an area like the Snake River Plain can have high temperatures of 38°C (100°F) during a summer day, and then have lows of 5–10°C (41–50°F). At the same time, Washington D.C., which is much more humid, has temperature variations of only 8°C (14°F); urban Hong Kong has a diurnal temperature range of little more than 4°C (7.2°F).

While the National Park Service claimed that the world record is a variation of 102°F (56.7°C) (from 46°F or 7.8°C to −56°F or −48.9°C) in Browning, Montana in 1916, the Montana Department of Environmental Quality claimed that Loma, Montana also had a variation of 102°F (56.7°C) (from −54°F or −47.8°C to 48°F or 8.9°C) in 1972.

Viticulture

Diurnal temperature variation is of particular importance in viticulture. Wine regions situated in areas of high altitude experience the most dramatic swing in temperature variation during the course of a day. In grapes, this variation has the effect of producing high acid and high sugar content as the grapes' exposure to sunlight increases the ripening qualities while the sudden drop in temperature at night preserves the balance of natural acids in the grape.

References

- Harvey, Samantha (1 May 2006). "Weather, Weather, Everywhere?". NASA. Archived from the original on 8 August 2007. Retrieved 9 September 2007

- Zeilik, Michael A.; Gregory, Stephan A. (1998). Introductory Astronomy & Astrophysics (4th ed.). Saunders College Publishing. p. 67. ISBN 0-03-006228-4

- Knutson, Heather A.; Charbonneau, David; Allen, Lori E.; Fortney, Jonathan J. (2007). "A map of the day-night contrast of the extrasolar planet HD 189733b". Nature. 447 (7141): 183–6. Bibcode:2007Natur.447..183K. PMID 17495920. arXiv:0705.0993. doi:10.1038/nature05782

- Sheppard, S. S.; Jewitt, D.; Kleyna, J. (2005). "An Ultradeep Survey for Irregular Satellites of Uranus: Limits to Completeness". The Astronomical Journal. 129: 518. Bibcode:2005AJ....129..518S. arXiv:astro-ph/0410059. doi:10.1086/426329

- Brasseur, Guy (1984). Aeronomy of the Middle Atmosphere : Chemistry and Physics of the Stratosphere and Mesosphere. Springer. pp. xi. ISBN 978-94-009-6403-7

- Wheeling Jesuit University. Exploring the Environment: UV Menace. Archived August 30, 2007, at the Wayback Machine. Retrieved on 2007-06-01

- Ballester, Gilda E.; Sing, David K.; Herbert, Floyd (2007). "The signature of hot hydrogen in the atmosphere of the extrasolar planet HD 209458b". Nature. 445 (7127): 511–4. Bibcode:2007Natur.445..511B. PMID 17268463. doi:10.1038/nature05525

- Nagy, Andrew F.; Balogh, André; Thomas E. Cravens; Mendillo, Michael; Mueller-Woodarg, Ingo (2008). Comparative Aeronomy. Springer. pp. 1–2. ISBN 978-0-387-87824-9

- Dr. Hugh J. Christian and Melanie A. McCook. Lightning Detection From Space: A Lightning Primer. Archived April 30, 2008, at the Wayback Machine. Retrieved on 2008-04-17

- Harrington, Jason; Hansen, Brad M.; Luszcz, Statia H.; Seager, Sara (2006). "The phase-dependent infrared brightness of the extrasolar planet Andromeda b". Science. 314 (5799): 623–6. Bibcode:2006Sci...314..623H. PMID 17038587. arXiv:astro-ph/0610491. doi:10.1126/science.1133904

Principle Layers of the Earth's Atmosphere

The Earth's atmosphere can be divided into several layers. The five layers that exist are troposphere, stratosphere, mesosphere, thermosphere and exosphere. Tropopause is the boundary between the troposphere and the stratosphere. The major components of the Earth's atmosphere are discussed in this chapter.

Troposphere

The troposphere is the lowest portion of Earth's atmosphere, and is also where nearly all weather takes place. It contains approximately 75% of the atmosphere's mass and 99% of the total mass of water vapor and aerosols. The average depths of the troposphere are 20 km (12 mi) in the tropics, 17 km (11 mi) in the mid latitudes, and 7 km (4.3 mi) in the polar regions in winter. The lowest part of the troposphere, where friction with the Earth's surface influences air flow, is the planetary boundary layer. This layer is typically a few hundred meters to 2 km (1.2 mi) deep depending on the landform and time of day. Atop the troposphere is the tropopause, which is the border between the troposphere and stratosphere. The tropopause is an inversion layer, where the air temperature ceases to decrease with height and remains constant through its thickness.

The word *troposphere* derives from the Greek: *tropos* for "turn, turn toward, trope" and "-sphere" (as in, the Earth), reflecting the fact that rotational turbulent mixing plays an important role in the troposphere's structure and behaviour. Most of the phenomena associated with day-to-day weather occur in the troposphere.

Pressure and Temperature Structure

A view of Earth's troposphere from an airplane

Composition

By volume, dry air contains 78.09% nitrogen, 20.95% oxygen, 0.93% argon, 0.04% carbon dioxide, and small amounts of other gases. Air also contains a variable amount of water vapor. The chemical composition of the troposphere is essentially uniform, with the notable exception of water vapor. The source of water vapor is at the surface through the processes of evaporation. The temperature of the troposphere decreases with height, and saturation vapor pressure decreases strongly as temperature drops, so the amount of water vapor that can exist in the atmosphere decreases strongly with height. Thus the proportion of water vapor is normally greatest near the surface and decreases with height.

Pressure

The pressure of the atmosphere is maximum at sea level and decreases with altitude. This is because the atmosphere is very nearly in hydrostatic equilibrium, so that the pressure is equal to the weight of air above a given point. The change in pressure with height, therefore can be equated to the density with this hydrostatic equation:

$$\frac{dP}{dz} = -\rho g_n = -\frac{mPg_n}{RT}$$

where:

- g_n is the standard gravity

- ρ is the density

- z is the altitude

- P is the pressure

- R is the gas constant

- T is the thermodynamic (absolute) temperature

- m is the molar mass

Since temperature in principle also depends on altitude, one needs a second equation to determine the pressure as a function of height.

Temperature

The temperature of the troposphere generally decreases as altitude increases. The rate at which the temperature decreases, , is called the environmental lapse rate (ELR). The ELR is nothing more than the difference in temperature between the surface and the tropopause divided by the height. The ELR assumes that the air is perfectly still, i.e. that there is no mixing of the layers of air from vertical convection, nor winds that would create turbulence and hence mixing of the layers of air. The reason for this temperature difference is that the ground absorbs most of the sun's energy, which then heats the lower levels of the atmosphere with which it is in contact. Meanwhile, the radiation of heat at the top of the atmosphere results in the cooling of that part of the atmosphere.

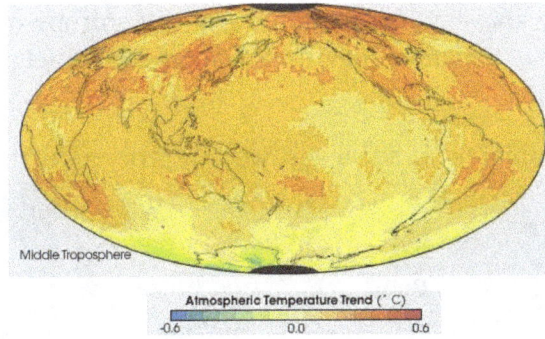

Atmospheric Temperature Trend (°C)

This image shows the temperature trend in the Middle Troposphere as measured by a series of satellite-based instruments between January 1979 and December 2005. The middle troposphere is centered around 5 kilometers above the surface. Oranges and yellows dominate the troposphere image, indicating that the air nearest the Earth's surface warmed during the period.

Environmental Lapse Rate (ELR)		
Altitude Region	**Lapse rate**	**Lapse Rate**
(m)	**(Kelvin/km)**	**(°F/1000 feet)**
0 – 11,000	−6.5	−3.57
11,000 – 20,000	0.0	0.0
20,000 – 32,000	1.0	0.55
32,000 – 47,000	2.8	1.54
47,000 – 51,000	0.0	0.0
51,000 – 71,000	−2.8	−1.54
71,000 – 85,000	−2.0	−1.09

The ELR assumes the atmosphere is still, but as air is heated it becomes buoyant and rises. The dry adiabatic lapse rate accounts for the effect of the expansion of dry air as it rises in the atmosphere and wet adiabatic lapse rates includes the effect of the condensation of water vapor on the lapse rate.

When a parcel of air rises, it expands, because the pressure is lower at higher altitudes. As the air parcel expands, it pushes the surrounding air outward, transferring energy in the form of work from that parcel to the atmosphere. As energy transfer to a parcel of air by way of heat is very slow, it is assumed to not exchange energy by way of heat with the environment. Such a process is called an adiabatic process (no energy transfer by way of heat). Since the rising parcel of air is losing energy as it does work on the surrounding atmosphere and no energy is transferred into it as heat from the atmosphere to make up for the loss, the parcel of air is losing energy, which manifests itself as a decrease in the temperature of the air parcel. The reverse, of course, will be true for a parcel of air that is sinking and is being compressed.

Since the process of compression and expansion of an air parcel can be considered reversible and no energy is transferred into or out of the parcel, such a process is considered isentropic, meaning that there is no change in entropy as the air parcel rises and falls, $dS = 0$. Since the heat exchanged $dQ = 0$ is related to the entropy change dS by $dQ = TdS$, the equation governing the temperature as a function of height for a thoroughly mixed atmosphere is

$$\frac{dS}{dz} = 0$$

where S is the entropy. The above equation states that the entropy of the atmosphere does not change with height. The rate at which temperature decreases with height under such conditions is called the adiabatic lapse rate.

For *dry* air, which is approximately an ideal gas, we can proceed further. The adiabatic equation for an ideal gas is

$$p(z)T(z)^{-\frac{\gamma}{\gamma-1}} = constant$$

where γ is the heat capacity ratio (γ =7/5, for air). Combining with the equation for the pressure, one arrives at the dry adiabatic lapse rate,

$$\frac{dT}{dz} = -\frac{mg}{R}\frac{\gamma-1}{\gamma} = -9.8\,^{\circ}C/km$$

If the air contains water vapor, then cooling of the air can cause the water to condense, and the behavior is no longer that of an ideal gas. If the air is at the saturated vapor pressure, then the rate at which temperature drops with height is called the saturated adiabatic lapse rate. More generally, the actual rate at which the temperature drops with altitude is called the environmental lapse rate. In the troposphere, the average environmental lapse rate is a drop of about 6.5°C for every 1 km (1,000 meters) in increased height.

The environmental lapse rate (the actual rate at which temperature drops with height, dT/dz) is not usually equal to the adiabatic lapse rate (or correspondingly, $dS/dz \neq 0$). If the upper air is warmer than predicted by the adiabatic lapse rate ($dS/dz > 0$), then when a parcel of air rises and expands, it will arrive at the new height at a lower temperature than its surroundings. In this case, the air parcel is denser than its surroundings, so it sinks back to its original height, and the air is stable against being lifted. If, on the contrary, the upper air is cooler than predicted by the adiabatic lapse rate, then when the air parcel rises to its new height it will have a higher temperature and a lower density than its surroundings, and will continue to accelerate upward.

The troposphere is heated from below by latent heat, longwave radiation, and sensible heat. Surplus heating and vertical expansion of the troposphere occurs in the tropics. At middle latitudes, tropospheric temperatures decrease from an average of 15°C at sea level to about −55°C at the tropopause. At the poles, tropospheric temperature only decreases from an average of 0°C at sea level to about −45°C at the tropopause. At the equator, tropospheric temperatures decrease from an average of 20°C at sea level to about −70 to −75°C at the tropopause. The troposphere is thinner at the poles and thicker at the equator. The average thickness of the tropical tropopause is roughly 7 kilometers greater than the average tropopause thickness at the poles.

Tropopause

The tropopause is the boundary region between the troposphere and the stratosphere.

Measuring the temperature change with height through the troposphere and the stratosphere identifies the location of the tropopause. In the troposphere, temperature decreases with altitude. In the stratosphere, however, the temperature remains constant for a while and then increases with altitude. The region of the atmosphere where the lapse rate changes from positive (in the troposphere) to negative (in the stratosphere), is defined as the tropopause. Thus, the tropopause is an inversion layer, and there is little mixing between the two layers of the atmosphere.

Atmospheric Flow

The flow of the atmosphere generally moves in a west to east direction. This, however, can often become interrupted, creating a more north to south or south to north flow. These scenarios are often described in meteorology as zonal or meridional. These terms, however, tend to be used in reference to localised areas of atmosphere (at a synoptic scale). A fuller explanation of the flow of atmosphere around the Earth as a whole can be found in the three-cell model.

Zonal Flow

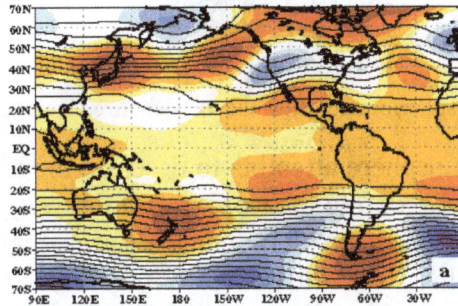

A zonal flow regime. Note the dominant west-to-east flow
as shown in the 500 hPa height pattern.

A zonal flow regime is the meteorological term meaning that the general flow pattern is west to east along the Earth's latitude lines, with weak shortwaves embedded in the flow. The use of the word "zone" refers to the flow being along the Earth's latitudinal "zones". This pattern can buckle and thus become a meridional flow.

Meridional Flow

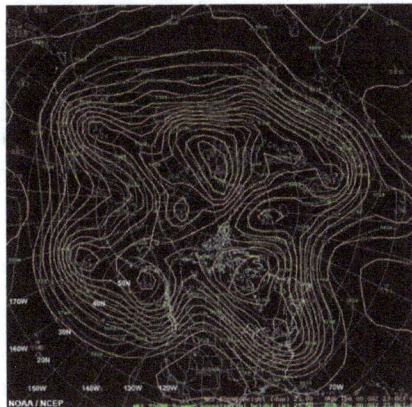

Meridional Flow pattern of October 23, 2003. Note the amplified
troughs and ridges in this 500 hPa height pattern.

When the zonal flow buckles, the atmosphere can flow in a more longitudinal (or meridional) direction, and thus the term "meridional flow" arises. Meridional flow patterns feature strong, amplified troughs of low pressure and ridges of high pressure, with more north-south flow in the general pattern than west-to-east flow.

Three-cell model

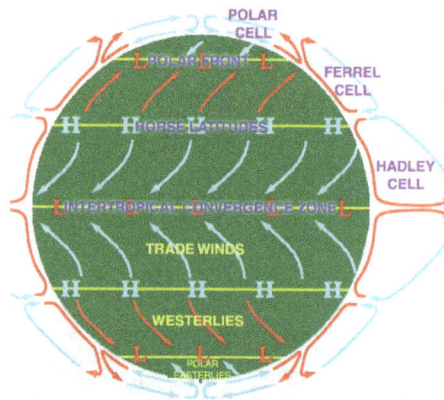

Atmospheric circulation shown with three large cells.

The three cells model of the atmosphere attempts to describe the actual flow of the Earth's atmosphere as a whole. It divides the Earth into the tropical (Hadley cell), mid latitude (Ferrel cell), and polar (polar cell) regions, to describe energy flow and global atmospheric circulation (mass flow). Its fundamental principle is that of balance – the energy that the Earth absorbs from the sun each year is equal to that which it loses to space by radiation. This overall Earth energy balance, however, does not apply in each latitude due to the varying strength of the sun in each "cell" as a result of the tilt of the Earth's axis in relation to its orbit. The result is a circulation of the atmosphere that transports warm air poleward from the tropics and cold air equatorward from the poles. The effect of the three cells is the tendency to even out the heat and moisture in the Earth's atmosphere around the planet.

Synoptic Scale Observations and Concepts

Forcing Term by Meteorologists

Forcing is a term used by meteorologists to describe the situation where a change or an event in one part of the atmosphere causes a strengthening change in another part of the atmosphere. It is usually used to describe connections between upper, middle or lower levels (such as upper-level divergence causing lower level convergence in cyclone formation), but also be to describe such connections over lateral distance rather than height alone. In some respects, teleconnections could be considered a type of forcing.

Divergence and Convergence

An area of convergence is one in which the total mass of air is increasing with time, resulting in an increase in pressure at locations below the convergence level (recall that atmospheric pressure is just the total weight of air above a given point). Divergence is the opposite of convergence – an area where the total mass of air is decreasing with time, resulting in falling pressure in regions below the area of divergence. Where divergence is occurring in the upper atmosphere, there will be air coming in to try to balance the net loss of mass (this is called the principle of mass conservation), and there is a resulting upward motion (positive vertical velocity). Another way to state this is to say that regions of upper air divergence are conducive to lower level convergence, cyclone formation, and positive vertical velocity. Therefore, identifying regions of upper air divergence is an important step in forecasting the formation of a surface low pressure area.

Tropopause

The tropopause is the boundary in the Earth's atmosphere between the troposphere and the stratosphere. It is a thermodynamic gradient stratification layer, marking the end of troposphere.

Definition

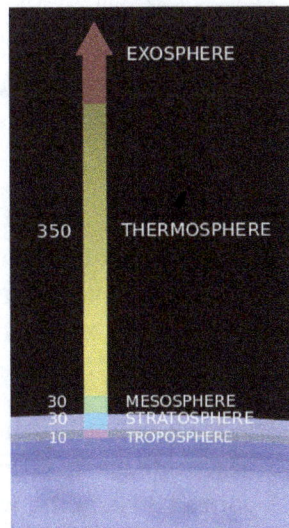

Schematic showing the different layers of the atmosphere (not to scale). The tropopause is located between the troposphere and the stratosphere.

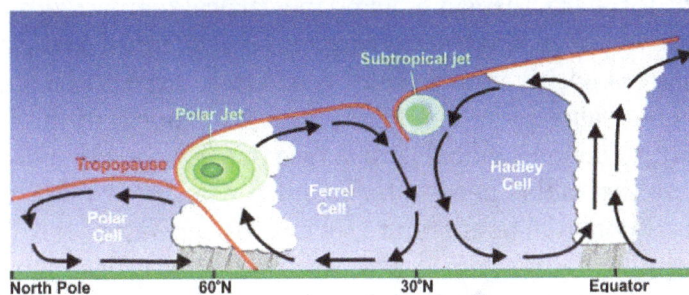

The tropopause lies higher in the tropics than at the poles.

Going upward from the surface, it is the point where air ceases to cool with height, and becomes almost completely dry. More formally, the tropopause is the region of the atmosphere where the environmental lapse rate changes from positive, as it behaves in the troposphere, to the stratospheric negative one. Following is the exact definition used by the World Meteorological Organization:

The boundary between the troposphere and the stratosphere, where an abrupt change in lapse rate usually occurs. It is defined as the lowest level at which the lapse rate decreases to 2 °C/km or less, provided that the average lapse rate between this level and all higher levels within 2 km does not exceed 2 °C/km.

The tropopause as defined above renders as a first-order discontinuity surface, that is, *temperature* as a function of height varies continuously through the atmosphere but the *temperature gradient* does not.

Location

The troposphere is one of the lowest layers of the Earth's atmosphere; it is located right above the planetary boundary layer, and is the layer in which most weather phenomena take place. The troposphere extends upwards from right above the boundary layer, and ranges in height from an average of 9 km (5.6 mi; 30,000 ft) at the poles, to 17 km (11 mi; 56,000 ft) at the Equator. In the absence of inversions and not considering moisture, the temperature lapse rate for this layer is 6.5°C per kilometer, on average, according to the *U.S. Standard Atmosphere*. A measurement of both the tropospheric and the stratospheric lapse rates helps identifying the location of the tropopause, since temperature increases with height in the stratosphere, and hence the lapse rate becomes negative. The tropopause location coincides with the lowest point at which the lapse rate falls below a prescribed threshold.

Since the tropopause responds to the average temperature of the entire layer that lies underneath it, it is at its peak levels over the Equator, and reaches minimum heights over the poles. On account of this, the coolest layer in the atmosphere lies at about 17 km over the equator. Due to the variation in starting height, the tropopause extremes are referred to as the equatorial tropopause and the polar tropopause.

Given that the lapse rate is not a conservative quantity when the tropopause is considered for stratosphere-troposphere exchanges studies, there exists an alternative definition named *dynamic tropopause*. It is formed with the aid of potential vorticity, which is defined as the product of the isentropic density, i.e. the density that arises from using potential temperature as the vertical co-ordinate, and the absolute vorticity, given that this quantity attains quite different values for the troposphere and the stratosphere. Instead of using the vertical temperature gradient as the defining variable, the dynamic tropopause surface is expressed in *potential vorticity units* (PVU). Given that the absolute vorticity is positive in the Northern Hemisphere and negative in the Southern Hemisphere, the threshold value should be taken as positive north of the Equator and negative south of it. Theoretically, to define a global tropopause in this way, the two surfaces arising from the positive and negative thresholds need to be matched near the equator using another type of surface such as a constant potential temperature surface. Nevertheless, the dynamic tropopause is useless at equatorial latitudes because the isentropes are almost vertical. For the extratropical tropopause in the Northern Hemisphere the WMO established a value of 1.5 PVU, but greater values ranging between 2 and 3.5 PVU have been traditionally used.

It is also possible to define the tropopause in terms of chemical composition. For example, the lower stratosphere has much higher ozone concentrations than the upper troposphere, but much lower water vapor concentrations, so appropriate cutoffs can be used.

Phenomena

The tropopause is not a "hard" boundary. Vigorous thunderstorms, for example, particularly those of tropical origin, will overshoot into the lower stratosphere and undergo a brief (hour-order or less) low-frequency vertical oscillation. Such oscillation sets up a low-frequency atmospheric gravity wave capable of affecting both atmospheric and oceanic currents in the region.

Most commercial aircraft are flown in the lower stratosphere, just above the tropopause, where clouds are usually absent, as also are significant weather perturbations.

Stratosphere

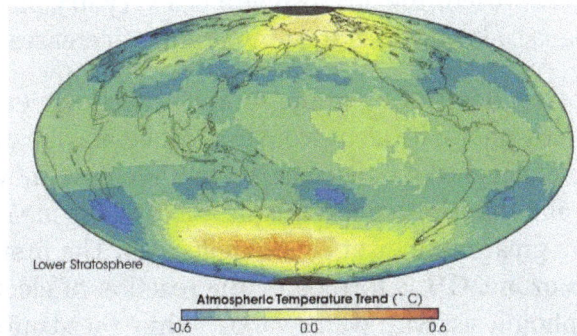

Lower Stratosphere

Atmospheric Temperature Trend (˚C)

-0.6   0.0   0.6

This image shows the temperature trend in the lower stratosphere as measured by a series of satellite-based instruments between January 1979 and December 2005. The lower stratosphere is centered around 63 kilometers above Earth's surface. The stratosphere image is dominated by blues and greens, which indicates a cooling over time.

The stratosphere is the second major layer of Earth's atmosphere, just above the troposphere, and below the mesosphere. About 20% of the atmosphere's mass is contained in the stratosphere. The stratosphere is stratified in temperature, with warmer layers higher and cooler layers closer to the Earth. The increase of temperature with altitude is a result of the absorption of the Sun's ultraviolet radiation by the ozone. This is in contrast to the troposphere, near the Earth's surface, where temperatures decreases with altitude. The border between the troposphere and stratosphere, the tropopause, marks where this temperature inversion begins. Near the equator, the stratosphere starts at 18 km (59,000 ft; 11 mi); at mid latitudes, it starts at 10–13 km (33,000–43,000 ft; 6.2–8.1 mi) and ends at 50 km (160,000 ft; 31 mi); at the poles, it starts at about 8 km (26,000 ft; 5.0 mi). Temperatures vary within the stratosphere with the seasons, in particular with the polar night (winter). The greatest variation of temperature takes place over the poles in the lower stratosphere; those variations are largely steady at lower latitudes and higher altitudes.

Ozone and Temperature

The mechanism describing the formation of the ozone layer was described by British mathematician Sydney Chapman in 1930. Molecular oxygen absorbs high energy sunlight in the UV-C region, at wavelengths shorter than about 240 nm. The oxygen atoms produced combine with molecular oxygen to form ozone. Ozone in turn is photolysed much more rapidly than molecular oxygen as it has a stronger absorption that occurs at longer wavelengths, where the solar emission is more intense. Ozone (O_3) photolysis produces O and O_2. The oxygen atom product combines with atmospheric molecular oxygen to reform O_3, releasing heat. The rapid photolysis and reformation of ozone heats the stratosphere resulting in a temperature inversion. This increase of temperature with altitude is characteristic of the stratosphere; its resistance to vertical mixing means that it is stratified. Within the stratosphere temperatures increase with altitude; the top of the stratosphere has a temperature of about 270 K (−3°C or 26.6°F). This vertical stratification, with warmer layers above and cooler layers below, makes the stratosphere dynamically stable: there is no regular convection and associated turbulence in this part of the atmosphere. However, exceptionally energetic convection processes, such as volcanic eruption columns and overshooting tops in severe supercell thunderstorms, may carry convection into the stratosphere on a very local and temporary basis. Overall the attenuation of solar UV at wavelengths that damage DNA by the ozone layer allows life to exist on the surface of the planet outside of the ocean. All air entering the stratosphere must pass through the tropopause, the temperature minimum that divides the troposphere and stratosphere. The rising air is literally freeze dried; the stratosphere is a very dry place. The top of the stratosphere is called the stratopause, above which the temperature decreases with height.

Sydney Chapman gave a correct description of the source of stratospheric ozone and its ability to generate heat within the stratosphere; he also wrote that ozone may be destroyed by reacting with atomic oxygen, making two molecules of molecular oxygen. We now know that there are additional ozone loss mechanisms, and that these mechanisms are catalytic meaning that a small amount of the catalyst can destroy a great number of ozone molecules. The first is due to the reaction of hydroxyl radicals OH· with ozone. OH is formed by the reaction of electronically excited oxygen atoms produced by ozone photolysis, with water vapor. While the stratosphere is dry, additional water vapor is produced in situ by the photochemical oxidation of methane (CH_4). The HO_2 radical produced by the reaction of OH with O_3 is recycled to OH by reaction with oxygen atoms or ozone. In addition, solar proton events can significantly effect ozone levels via Radiolysis with the subsequent formation of OH . Laughing gas or Nitrous oxide (N_2O) is produced by biological activity at the surface and is oxidised to NO in the stratosphere; the so-called NOx radical cycles also deplete stratospheric ozone. Finally, chlorofluorocarbon molecules are photolysed in the stratosphere releasing chlorine atoms that react with ozone giving ClO and O_2. The chlorine atoms are recycled when ClO reacts with O in the upper stratosphere, or when ClO reacts with itself in the chemistry of the Antarctic ozone hole. Paul J. Crutzen, Mario J. Molina and F. Sherwood Rowland were awarded the Nobel Prize in Chemistry in 1995 for their work describing the formation and decomposition of stratospheric ozone.

Aircraft Flight

Commercial airliners typically cruise at altitudes of 9–12 km (30,000–39,000 ft) which is in the lower reaches of the stratosphere in temperate latitudes. This optimizes fuel efficiency, mostly due to the low temperatures encountered near the tropopause and low air density, reducing parasitic drag on the airframe. Stated another way, it allows the airliner to fly faster for the same amount of drag. It also allows them to stay above the turbulent weather of the troposphere.

The Concorde aircraft cruised at mach 2 at about 18,000 m (59,000 ft), and the SR-71 cruised at mach 3 at 26,000 m (85,000 ft), all within the stratosphere.

Because the temperature in the tropopause and lower stratosphere is largely constant with increasing altitude, very little convection and its resultant turbulence occurs there. Most turbulence at this altitude is caused by variations in the jet stream and other local wind shears, although areas of significant convective activity (thunderstorms) in the troposphere below may produce turbulence as a result of convective overshoot.

On October 24, 2014, Alan Eustace became the record holder for reaching the altitude record for a manned balloon at 135,890 ft (41,419 m). Dr Eustace also broke the world records for vertical speed reached with a peak velocity of 1,321 km/h (822 mph) and total freefall distance of 123,414 ft (37,617 m) – lasting four minutes and 27 seconds.

Circulation and Mixing

The stratosphere is a region of intense interactions among radiative, dynamical, and chemical processes, in which the horizontal mixing of gaseous components proceeds much more rapidly than does vertical mixing. The overall circulation of the stratosphere is termed as Brewer-Dobson circulation, which is a single celled circulation, spanning from the tropics up to the poles, consisting of the tropical upwelling of air from the tropical troposphere and the extra tropical downwelling of air. Stratospheric Circulation is a pre-dominantly wave-driven circulation in that the tropical upwelling is induced by the wave force by the westward propagating Rossby Waves, in a phenomenon called Rossby-Wave pumping.

An interesting feature of stratospheric circulation is the quasi-biennial oscillation (QBO) in the tropical latitudes, which is driven by gravity waves that are convectively generated in the troposphere. The QBO induces a secondary circulation that is important for the global stratospheric transport of tracers, such as ozone or water vapor.

Another large-scale feature that significantly influences Stratospheric Circulation is the breaking planetary waves resulting in intense quasi-horizontal mixing in the midlatitudes. This breaking is much more pronounced in the winter hemisphere where this region is called the surf zone. This breaking is caused due to a highly non-linear interaction between the vertically propagating planetary waves and the isolated high potential vorticity region known as Polar Vortex. The resultant breaking causes large scale mixing of air and other trace gases throughout the midlatitude surf zone. The timescale of this rapid mixing is much smaller than the much slower timescales of upwelling in the tropics and downwelling in the extratropics.

During northern hemispheric winters, sudden stratospheric warmings, caused by the absorption of Rossby waves in the stratosphere, can be observed in approximately half of winters when easterly winds develop in the stratosphere. These events often precede unusual winter weather and may even be responsible for the cold European winters of the 1960s.

Stratospheric warming of the polar vortex results in its weakening. When the vortex is strong, it keeps the cold, high pressure air masses "contained" in the arctic; when the vortex weakens, air masses move equatorward, and results in rapid changes of weather in the mid latitudes.

Life

Bacteria

Bacterial life survives in the stratosphere, making it a part of the biosphere. In 2001 an Indian experiment, involving a high-altitude balloon, was carried out at a height of 41 kilometres and a sample of dust was collected with bacterial material inside.

Birds

Some bird species have been reported to fly at the lower levels of the stratosphere. On November 29, 1973, a Rüppell's vulture was ingested into a jet engine 11,278 m (37,000 ft) above the Ivory Coast, and bar-headed geese reportedly overfly Mount Everest's summit, which is 8,848 m (29,029 ft).

Discovery

Léon Teisserenc de Bort from France and Richard Assmann from Germany, in separate publications and following years of observations, announced the discovery of an isothermal layer at around 11–14 km, which is the base of the lower stratosphere. This was based on temperature profiles from unmanned instrumented balloons.

Mesosphere

The mesosphere is the layer of the Earth's atmosphere that is directly above the stratosphere and directly below the mesopause. In the mesosphere, temperature decreases as the altitude increases. The upper boundary of the mesosphere is the mesopause, which can be the coldest naturally occurring place on Earth with temperatures below $-143°C$ ($-225°F$; 130 K). The exact upper and lower boundaries of the mesosphere vary with latitude and with season, but the lower boundary of the mesosphere is usually located at heights of about 50 kilometres (160,000 ft; 31 mi) above the Earth's surface and the mesopause is usually at heights near 100 kilometres (62 mi), except at middle and high latitudes in summer where it descends to heights of about 85 kilometres (53 mi; 279,000 ft).

The stratosphere, mesosphere and lowest part of the thermosphere are collectively referred to as the "middle atmosphere", which spans heights from approximately 10 kilometres (33,000 ft; 6.2 mi) to 100 kilometres (62 mi; 330,000 ft). The mesopause, at an altitude of 80–90 km (50–56 mi), separates the mesosphere from the thermosphere—the second-outermost layer of the Earth's atmosphere. This is also around the same altitude as the turbopause, below which different chemical species are well mixed due to turbulent eddies. Above this level the atmosphere becomes non-uniform; the scale heights of different chemical species differ by their molecular masses.

Temperature

Within the mesosphere, temperature decreases with increasing height, due to decreasing solar heating and increasing cooling by CO_2 radiative emission. The top of the mesosphere, called the mesopause, is the coldest part of Earth's atmosphere. Temperatures in the upper mesosphere fall as low as $-101°C$ (172 K; $-150°F$), varying according to latitude and season.

Dynamic Features

The main dynamic features in this region are strong zonal (East-West) winds, atmospheric tides, internal atmospheric gravity waves (commonly called "gravity waves"), and planetary waves. Most of these tides and waves start in the troposphere and lower stratosphere, and propagate to the mesosphere. In the mesosphere, gravity-wave amplitudes can become so large that the waves become unstable and dissipate. This dissipation deposits momentum into the mesosphere and largely drives global circulation.

Noctilucent clouds are located in the mesosphere. The upper mesosphere is also the region of the ionosphere known as the *D layer*. The D layer is only present during the day, when some ionization occurs with nitric oxide being ionized by Lyman series-alpha hydrogen radiation. The ionization is so weak that when night falls, and the source of ionization is removed, the free electron and ion form back into a neutral molecule. The mesosphere has been called the "ignorosphere" because it is poorly studied relative to the stratosphere (which can be accessed with high-altitude balloons) and the thermosphere (in which satellites can orbit).

A 5 km (3.1 mi; 16,000 ft) deep sodium layer is located between 80–105 km (50–65 mi; 262,000–344,000 ft). Made of unbound, non-ionized atoms of sodium, the sodium layer radiates weakly to contribute to the airglow. The sodium has an average concentration of 400,000 atoms per cubic centimeter. This band is regularly replenished by sodium sublimating from incoming meteors. Astronomers have begun utilizing this sodium band to create "guide stars" as part of the adaptive optical correction process used to produce ultra-sharp ground-based observations.

Millions of meteors enter the Earth's atmosphere, averaging 40 tons per year.

Uncertainties

The mesosphere lies above altitude records for aircraft, while only the lowest few kilometers are accessible to modern balloons, for which the altitude record is 53.0 km. Meanwhile, the mesosphere is below the minimum altitude for orbital spacecraft due to high atmospheric drag. It has only been accessed through the use of sounding rockets, which are only capable of taking mesospheric measurements for a few minutes per mission. As a result, it is the least-understood part of the atmosphere. The presence of red sprites and blue jets (electrical discharges or lightning within the lower mesosphere), noctilucent clouds, and density shears within this poorly understood layer are of current scientific interest.

Thermosphere

The thermosphere is the layer of the Earth's atmosphere directly above the mesosphere. The exosphere is above that but is a minor layer of the atmosphere. Within this layer of the atmosphere, ultraviolet radiation causes photoionization/photodissociation of molecules, creating ions in the ionosphere. Taking its name from the Greek (pronounced *thermos*) meaning heat, the thermosphere begins about 85 kilometres (53 mi) above the Earth. At these high altitudes, the residual atmospheric gases sort into strata according to molecular mass. Thermospheric temperatures increase with altitude due to absorption of highly energetic solar radiation. Temperatures are highly dependent on solar activity, and can rise to 2,000°C (3,630°F). Radiation causes the atmosphere particles in this layer to become electrically charged, enabling radio waves to be refracted and thus be received beyond the horizon. In the exosphere, beginning at 500 to 1,000 kilometres (310 to

620 mi) above the Earth's surface, the atmosphere turns into space, although by the criteria set for the definition of the Karman line, the thermosphere itself is part of space.

The highly diluted gas in this layer can reach 2,500°C (4,530°F) during the day. Even though the temperature is so high, one would not feel warm in the thermosphere, because it is so near vacuum that there is not enough contact with the few atoms of gas to transfer much heat. A normal thermometer might be significantly below 0°C (32°F), at least at night, because the energy lost by thermal radiation would exceed the energy acquired from the atmospheric gas by direct contact. In the anacoustic zone above 160 kilometres (99 mi), the density is so low that molecular interactions are too infrequent to permit the transmission of sound.

The dynamics of the thermosphere are dominated by atmospheric tides, which are driven by the very significant diurnal heating. Atmospheric waves dissipate above this level because of collisions between the neutral gas and the ionospheric plasma.

The International Space Station orbits the Earth within the middle of the thermosphere, between 330 and 435 kilometres (205 and 270 mi).

Neutral Gas Constituents

It is convenient to separate the atmospheric regions according to the two temperature minima at about 12 km altitude (the tropopause) and at about 85 km (the mesopause) (Figure below). The thermosphere (or the upper atmosphere) is the height region above 85 km, while the region between the tropopause and the mesopause is the middle atmosphere (stratosphere and mesosphere) where absorption of solar UV radiation generates the temperature maximum near 45 km altitude and causes the ozone layer.

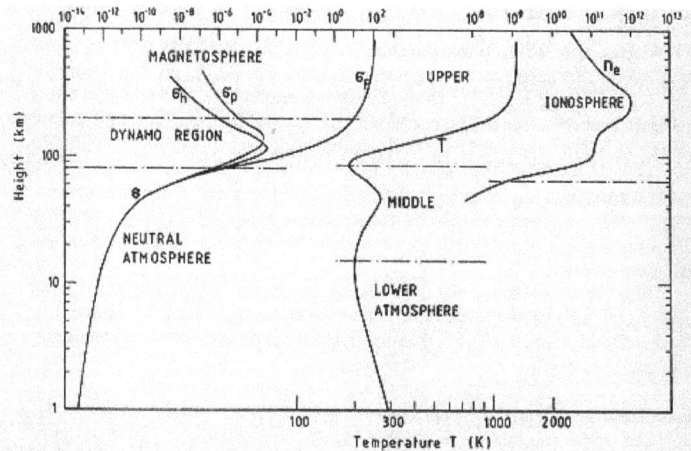

Nomenclature of atmospheric regions based on the profiles of electric conductivity (left), temperature (middle), and electron number density in m^{-3}

The density of the Earth's atmosphere decreases nearly exponentially with altitude. The total mass of the atmosphere is $M = \rho_A H \simeq 1\,kg/cm^2$ within a column of one square centimeter above the ground (with $\rho_A = 1.29\,kg/m^3$ the atmospheric density on the ground at $z = 0$ m altitude, and $H \simeq 8$ km the average atmospheric scale height). 80% of that mass is concentrated within the troposphere. The mass of the thermosphere above about 85 km is only 0.002% of the total mass. Therefore, no significant energetic feedback from the thermosphere to the lower atmospheric regions can be expected.

Turbulence causes the air within the lower atmospheric regions below the turbopause at about 110 km to be a mixture of gases that does not change its composition. Its mean molecular weight is 29 g/mol with molecular oxygen (O_2) and nitrogen (N_2) as the two dominant constituents. Above the turbopause, however, diffusive separation of the various constituents is significant, so that each constituent follows its own barometric height structure with a scale height inversely proportional to its molecular weight. The lighter constituents atomic oxygen (O), helium (He), and hydrogen (H) successively dominate above about 200 km altitude and vary with geographic location, time, and solar activity. The ratio N_2/O which is a measure of the electron density at the ionospheric F region is highly affected by these variations. These changes follow from the diffusion of the minor constituents through the major gas component during dynamic processes.

The thermosphere contains an appreciable concentration of elemental sodium located in a 10-km thick band that occurs at the edge of the mesosphere, 80 to 100 km above Earth's surface. The sodium has an average concentration of 400,000 atoms per cubic centimeter. This band is regularly replenished by sodium sublimating from incoming meteors. Astronomers have begun utilizing this sodium band to create "guide stars" as part of the optical correction process in producing ultra-sharp ground-based observations.

Energy Input

Energy Budget

The thermospheric temperature can be determined from density observations as well as from direct satellite measurements. The temperature vs. altitude z in figure can be simulated by the so-called Bates profile:

(1) $$T = T_\infty - (T_\infty - T_0)\exp\{-s(z - z_0)\}$$

with T_∞ the exospheric temperature above about 400 km altitude, T_0 = 355 K, and z_0 = 120 km reference temperature and height, and s an empirical parameter depending on T_∞ and decreasing with T_∞. That formula is derived from a simple equation of heat conduction. One estimates a total heat input of $q_0 \approx 0.8$ to 1.6 mW/m² above z_0 = 120 km altitude. In order to obtain equilibrium conditions, that heat input q_0 above z_0 is lost to the lower atmospheric regions by heat conduction.

The exospheric temperature T_∞ is a fair measurement of the solar XUV radiation. Since solar radio emission F at 10.7 cm wavelength is a good indicator of solar activity, one can apply the empirical formula for quiet magnetospheric conditions.

(2) $$T_\infty \simeq 500 + 3.4F_0$$

with T_∞ in K, F_0 in 10^{-2} W m⁻² Hz⁻¹ (the Covington index) a value of F averaged over several solar cycles. The Covington index varies typically between 70 and 250 during a solar cycle, and never drops below about 50. Thus, T_∞ varies between about 740 and 1350 K. During very quiet magnetospheric conditions, the still continuously flowing magnetospheric energy input contributes by about 250 K to the residual temperature of 500 K in eq.(2). The rest of 250 K in eq.(2) can be attributed to atmospheric waves generated within the troposphere and dissipated within the lower thermosphere.

Solar XUV Radiation

The solar X-ray and extreme ultraviolet radiation (XUV) at wavelengths < 170 nm is almost completely absorbed within the thermosphere. This radiation causes the various ionospheric layers as well as a temperature increase at these heights (Figure above). While the solar visible light (380 to 780 nm) is nearly constant with a variability of not more than about 0.1% of the solar constant, the solar XUV radiation is highly variable in time and space. For instance, X-ray bursts associated with solar flares can dramatically increase their intensity over preflare levels by many orders of magnitude over a time span of tens of minutes. In the extreme ultraviolet, the Lyman α line at 121.6 nm represents an important source of ionization and dissociation at ionospheric D layer heights. During quiet periods of solar activity, it alone contains more energy than the rest of the XUV spectrum. Quasi-periodic changes of the order of 100% or greater, with periods with period of 27 days and 11 years, belong to the prominent variations of solar XUV radiation. However, irregular fluctuations over all time scales are present all the time. During low solar activity, about half of the total energy input into the thermosphere is thought to be solar XUV radiation. Evidently, that solar XUV energy input occurs only during daytime conditions, maximizing at the equator during equinox.

Solar Wind

A second source of energy input into the thermosphere is solar wind energy which is transferred to the magnetosphere by mechanisms that are not well understood. One possible way to transfer energy is via a hydrodynamic dynamo process. Solar wind particles penetrate into the polar regions of the magnetosphere where the geomagnetic field lines are essentially vertically directed. An electric field is generated, directed from dawn to dusk. Along the last closed geomagnetic field lines with their footpoints within the auroral zones, field aligned electric currents can flow into the ionospheric dynamo region where they are closed by electric Pedersen and Hall currents. Ohmic losses of the Pedersen currents heat the lower thermosphere. In addition, penetration of high energetic particles from the magnetosphere into the auroral regions enhance drastically the electric conductivity, further increasing the electric currents and thus Joule heating. During quiet magnetospheric activity, the magnetosphere contributes perhaps by a quarter to the thermosphere's energy budget. This is about 250 K of the exospheric temperature in eq.(2). During very large activity, however, this heat input can increase substantially, by a factor of four or more. That solar wind input occurs mainly in the auroral regions during both day and night.

Atmospheric Waves

Two kinds of large-scale atmospheric waves within the lower atmosphere exist: internal waves with finite vertical wavelengths which can transport wave energy upward; and external waves with infinitely large wavelengths which cannot transport wave energy. Atmospheric gravity waves and most of the atmospheric tides generated within the troposphere belong to the internal waves. Their density amplitudes increase exponentially with height, so that at the mesopause these waves become turbulent and their energy is dissipated (similar to breaking of ocean waves at the coast), thus contributing to the heating of the thermosphere by about 250 K in eq.(2). On the other hand, the fundamental diurnal tide labelled (1, −2) which is most efficiently excited by solar irradiance is an external wave and plays only a marginal role within lower and middle atmosphere. However, at thermospheric altitudes, it becomes the predominant wave. It drives the electric Sq-current within the ionospheric dynamo region between about 100 and 200 km height.

Heating, predominately by tidal waves, occurs mainly at lower and middle latitudes. The variability of this heating depends on the meteorological conditions within troposphere and middle atmosphere, and may not exceed about 50%.

Dynamics

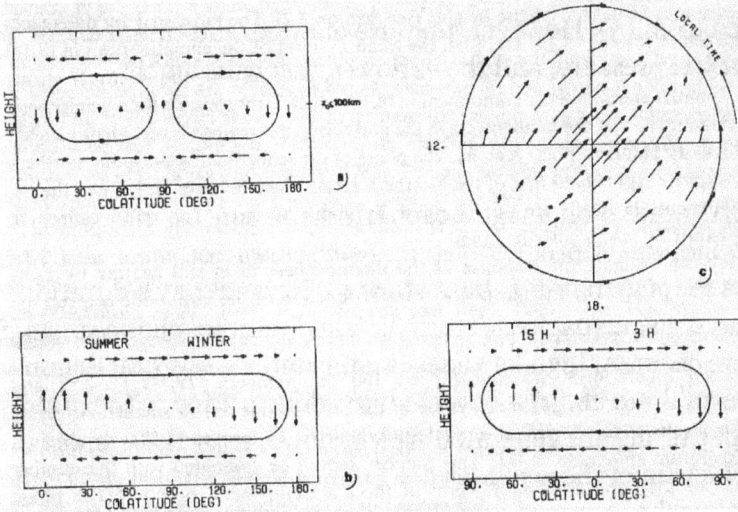

Schematic meridian-height cross-section of circulation of (a) symmetric wind component (P_2^0), (b) of antisymmetric wind component (P_1^0), and (d) of symmetric diurnal wind component (P_1^1) at 3 h and 15 h local time. Upper right pannel (c) shows the horizontal wind vectors of the diurnal component in the northern hemisphere depending on local time.

Within the thermosphere above about 150 km height, all atmospheric waves successively become external waves, and no significant vertical wave structure is visible. The atmospheric wave modes degenerate to the spherical functions P_n^m with m a meridional wave number and n the zonal wave number (m = 0: zonal mean flow; m = 1: diurnal tides; m = 2: semidiurnal tides; etc.). The thermosphere becomes a damped oscillator system with low-pass filter characteristics. This means that smaller-scale waves (greater numbers of (n,m)) and higher frequencies are suppressed in favor of large-scale waves and lower frequencies. If one considers very quiet magnetospheric disturbances and a constant mean exospheric temperature (averaged over the sphere), the observed temporal and spatial distribution of the exospheric temperature distribution can be described by a sum of spheric functions:

$$(3) \qquad T(\varphi, \lambda, t) = T_\infty \{1 + \Delta T_2^0 P_2^0(\varphi) + \Delta T_1^0 P_1^0(\varphi) \cos[\omega_a(t - t_a)] + \Delta T_1^1 P_1^1(\varphi) \cos(\tau - \tau_d) + \cdots\}$$

Here, it is φ latitude, λ longitude, and t time, ω_a the angular frequency of one year, ω_d the angular frequency of one solar day, and $\tau = \omega_d t + \lambda$ the local time. t_a = June 21 is the date of northern summer solstice, and τ_d = 15:00 is the local time of maximum diurnal temperature.

The first term in (3) on the right is the global mean of the exospheric temperature (of the order of 1000 K). The second term [with P_2^0 = 0.5(3 sin²(φ)−1)] represents the heat surplus at lower latitudes and a corresponding heat deficit at higher latitudes. A thermal wind system develops with wind toward the poles in the upper level and wind away from the poles in the lower level. The coefficient $\Delta T_2^0 \approx 0.004$ is small because Joule heating in the aurora regions compensates that heat surplus even during quiet magnetospheric conditions. During disturbed conditions, however, that term becomes dominant, changing sign so that now heat surplus is transported from the poles to the equator. The third term (with $P_1^o = \sin \varphi$) represents heat surplus on the summer hemisphere and is responsible for the transport of excess heat from the summer into the winter hemisphere. Its relative amplitude is of the order $\Delta T_1^o \simeq 0.13$. The fourth term (with $P_1^1(\varphi) = \cos \varphi$) is the dominant diurnal wave (the tidal mode (1,–2)). It is responsible for the transport of excess heat from the daytime hemisphere into the nighttime hemisphere. Its relative amplitude is $\Delta T_1^1 \simeq 0.15$, thus on the order of 150 K. Additional terms (e.g., semiannual, semidiurnal terms and higher order terms) must be added to eq.(3). However, they are of minor importance. Corresponding sums can be developed for density, pressure, and the various gas constituents.

Thermospheric Storms

In contrast to solar XUV radiation, magnetospheric disturbances, indicated on the ground by geomagnetic variations, show an unpredictable impulsive character, from short periodic disturbances of the order of hours to long-standing giant storms of several day's duration. The reaction of the thermosphere to a large magnetospheric storm is called thermospheric storm. Since the heat input into the thermosphere occurs at high latitudes (mainly into the auroral regions), the heat transport represented by the term P_2^o in eq.(3) is reversed. In addition, due to the impulsive form of the disturbance, higher-order terms are generated which, however, possess short decay times and thus quickly disappear. The sum of these modes determines the "travel time" of the disturbance to the lower latitudes, and thus the response time of the thermosphere with respect to the magnetospheric disturbance. Important for the development of an ionospheric storm is the increase of the ratio N_2/O during a thermospheric storm at middle and higher latitude. An increase of N_2 increases the loss process of the ionospheric plasma and causes therefore a decrease of the electron density within the ionospheric F-layer (negative ionospheric storm).

Exosphere

The exosphere (Ancient Greek outside, external, beyond", Ancient Greek *sphaîra* "sphere") is a thin, atmosphere-like volume surrounding a planet or natural satellite where molecules are gravitationally bound to that body, but where the density is too low for them to behave as a gas by colliding with each other. In the case of bodies with substantial atmospheres, such as Earth's atmosphere, the exosphere is the uppermost layer, where the atmosphere thins out and merges with interplanetary space. It is located directly above the thermosphere.

Mercury and several large moons, such as the Moon and the Galilean satellites of Jupiter, have exospheres without a denser atmosphere underneath, referred to as a surface boundary exosphere. Here, molecules are ejected on elliptic trajectories until they collide with the surface. Smaller bodies such as asteroids, in which the molecules emitted from the surface escape to space, are not considered to have exospheres.

Earth's Exosphere

The most common molecules within Earth's exosphere are those of the lightest atmospheric gasses. Hydrogen is present throughout the exosphere, with some helium, carbon dioxide, and atomic oxygen near its base. Because it can be difficult to define the boundary between the exosphere and outer space.

Lower Boundary

The lower boundary of the exosphere is called the *exobase*. It is also called *exopause* and *'critical altitude'* as this is the altitude where barometric conditions no longer apply. Atmospheric temperature becomes nearly a constant above this altitude. On Earth, the altitude of the exobase ranges from about 500 to 1,000 kilometres (310 to 620 mi) depending on solar activity.

The exobase can be defined in one of two ways:

If we define the exobase as the height at which upward-traveling molecules experience one collision on average, then at this position the mean free path of a molecule is equal to one pressure scale height. This is shown in the following. Consider a volume of air, with horizontal area A and height equal to the mean free path l, at pressure p and temperature T. For an ideal gas, the number of molecules contained in it is:

$$n = \frac{pAl}{RT}$$

where R is the universal gas constant. From the requirement that each molecule traveling upward undergoes on average one collision, the pressure is:

$$p = \frac{m_A n g}{A}$$

where m_A is the mean molecular mass of the gas. Solving these two equations gives:

$$l = \frac{RT}{m_A g}$$

which is the equation for the pressure scale height. As the pressure scale height is almost equal to the density scale height of the primary constituent, and because the Knudsen number is the ratio of mean free path and typical density fluctuation scale, this means that the exobase lies in the region where $\mathrm{Kn}(h_{EB}) \simeq 1$.

The fluctuation in the height of the exobase is important because this provides atmospheric drag on satellites, eventually causing them to fall from orbit if no action is taken to maintain the orbit.

Upper Boundary of Earth

In principle, the exosphere covers distances where particles are still gravitationally bound to Earth, i.e. particles still have ballistic orbits that will take them back towards Earth. The upper boundary of the exosphere can be defined as the distance at which the influence of solar radiation pressure on atomic hydrogen exceeds that of Earth's gravitational pull. This happens at half the distance to the Moon (the average distance between Earth and the Moon is 384,400 kilometres (238,900 mi)). The exosphere, observable from space as the geocorona, is seen to extend to at least

10,000 kilometres (6,200 mi) from Earth's surface. The exosphere is a transitional zone between Earth's atmosphere and space.

Moon's Exosphere

On 17 August 2015, based on studies with the Lunar Atmosphere and Dust Environment Explorer (LADEE) spacecraft, NASA scientists reported the detection of neon in the exosphere of the moon.

References

- Steigerwald, William (17 August 2015). "NASA's LADEE Spacecraft Finds Neon in Lunar Atmosphere". NASA. Retrieved 18 August 2015

- Gettelman, A.; Salby, M. L.; Sassi, F. (2002). "Distribution and influence of convection in the tropical tropopause region". Journal of Geophysical Research. American Geophysical Union. 107. Bibcode:2002JGRD..107.4080G. doi:10.1029/2001JD001048

- Jones, Daniel (2003) [1917], Peter Roach, James Hartmann and Jane Setter, eds., English Pronouncing Dictionary, Cambridge: Cambridge University Press, ISBN 3-12-539683-2

- Mesosphere (Wayback Machine Archive), Atmosphere, Climate & Environment Information ProgGFKDamme (UK Department for Environment, Food and Rural Affairs), archived from the original on 1 July 2010, retrieved 14 November 2011

- Hoinka, K. P. (1999). "Temperature, Humidity, and Wind at the Global Tropopause". Monthly Weather Review. American Meteorological Society. 127: 2248 – 2265. Bibcode:1999MWRv..127.2248H. doi:10.1175/1520 -0493(1999)127<2248:THAWAT>2.0.CO;2

- Hoinka, Klaus P. (December 1998). "Statistics of the Global Tropopause Pressure". Journal of Climate. American Meteorological Society (126): 3303 – 3325

- M. M. Woolfson. Time, Space, Stars & Man: The Story of the Big Bang. World Scientific; 2013. ISBN 978-1-84816-933-3. p. 388

- "NASA Sounding Rocket Program Overview". NASA Sounding Rocket Program. NASA. 24 July 2006. Retrieved 10 October 2006

- Hoskins, B. J.; McIntyre, M. E.; Robertson, A. W. (1985). "On the use and significance of isentropic potential vorticity maps". Quarterly Journal of the Royal Meteorological Society. 111: 877 – 946. Bibcode:1985Q-JRMS.111..877H. doi:10.1002/qj.49711147002

- Zängl, Günther; Hoinka, Klaus P. (15 July 2001). "The Tropopause in the Polar Regions". Journal of Climate. 14: 3117 – 3139. Bibcode:2001JCli...14.3117Z. doi:10.1175/1520-0442(2001)014<3117:ttitpr>2.0.co;2

- International Meteorological Vocabulary (2nd ed.). Geneva: Secretariat of the World Meteorological Organization. 1992. p. 636. ISBN 92-63-02182-1

- "Research on Balloon to Float over 50km Altitude". Institute of Space and Astronautical Science, JAXA. Retrieved 29 September 2011

- L. L. Pan; W. J. Randel; B. L. Gary; M. J. Mahoney; E. J. Hintsa (2004). "Definitions and sharpness of the extratropical tropopause: A trace gas perspective". Journal of Geophysical Research. 109: D23103. doi:10.1029/2004JD004982

3

Atmospheric Chemistry: An Integrated Study

The branch of atmospheric science that studies the chemical composition of the atmosphere of the Earth is known as atmospheric chemistry. The subject focuses on the changes faced by the atmosphere because of global warming. Some of the harms studied are ozone depletion, greenhouse gases, acid rain, etc. Atmospheric chemistry is best understood in confluence with the major topics listed in the following chapter.

Atmospheric Chemistry

Atmospheric chemistry is a branch of atmospheric science in which the chemistry of the Earth's atmosphere and that of other planets is studied. It is a multidisciplinary approach of research and draws on environmental chemistry, physics, meteorology, computer modeling, oceanography, geology and volcanology and other disciplines. Research is increasingly connected with other arenas of study such as climatology.

The composition and chemistry of the Earth's atmosphere is of importance for several reasons, but primarily because of the interactions between the atmosphere and living organisms. The composition of the Earth's atmosphere changes as result of natural processes such as volcano emissions, lightning and bombardment by solar particles from corona. It has also been changed by human activity and some of these changes are harmful to human health, crops and ecosystems. Examples of problems which have been addressed by atmospheric chemistry include acid rain, ozone depletion, photochemical smog, greenhouse gases and global warming. Atmospheric chemists seek to understand the causes of these problems, and by obtaining a theoretical understanding of them, allow possible solutions to be tested and the effects of changes in government policy evaluated.

Atmospheric Composition

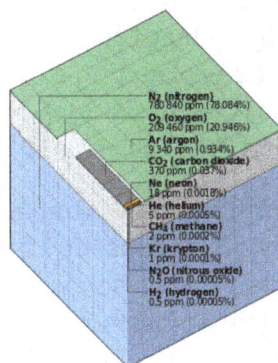

Visualisation of composition by volume of Earth's atmosphere. Water vapour is not included as it is highly variable. Each tiny cube (such as the one representing krypton) has one millionth of the volume of the entire block. Data is from NASA Langley.

Average composition of dry atmosphere (mole fractions)	
Gas	**per NASA**
Nitrogen, N_2	78.084%
Oxygen, O_2	20.946%
Minor constituents (mole fractions in ppm)	
Argon, Ar	9340
Carbon dioxide, CO_2	400
Neon, Ne	18.18
Helium, He	5.24
Methane, CH_4	1.7
Krypton, Kr	1.14
Hydrogen, H_2	0.55
Nitrous oxide, N_2O	0.5
Xenon, Xe	0.09
Nitrogen dioxide, NO_2	0.02
Water	
Water vapour	Highly variable; typically makes up about 1%

Notes: the concentration of CO_2 and CH_4 vary by season and location. The mean molecular mass of air is 28.97 g/mol. Ozone (O_3) is not included due to its high variability.

History

The ancient Greeks regarded air as one of the four elements. The first scientific studies of atmospheric composition began in the 18th century, as chemists such as Joseph Priestley, Antoine Lavoisier and Henry Cavendish made the first measurements of the composition of the atmosphere.

In the late 19th and early 20th centuries interest shifted towards trace constituents with very small concentrations. One particularly important discovery for atmospheric chemistry was the discovery of ozone by Christian Friedrich Schönbein in 1840.

In the 20th century atmospheric science moved on from studying the composition of air to a consideration of how the concentrations of trace gases in the atmosphere have changed over time and the chemical processes which create and destroy compounds in the air. Two particularly important examples of this were the explanation by Sydney Chapman and Gordon Dobson of how the ozone layer is created and maintained, and the explanation of photochemical smog by Arie Jan Haagen-Smit. Further studies on ozone issues led to the 1995 Nobel Prize in Chemistry award shared between Paul Crutzen, Mario Molina and Frank Sherwood Rowland.

In the 21st century the focus is now shifting again. Atmospheric chemistry is increasingly studied as one part of the Earth system. Instead of concentrating on atmospheric chemistry in isolation the focus is now on seeing it as one part of a single system with the rest of the atmosphere, biosphere and geosphere. An especially important driver for this is the links between chemistry and climate such as the effects of changing climate on the recovery of the ozone hole and vice versa but also interaction of the composition of the atmosphere with the oceans and terrestrial ecosystems.

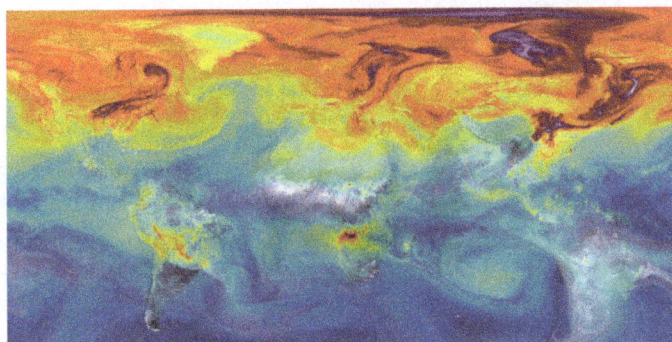

Carbon dioxide in Earth's atmosphere if *half* of global-warming emissions are *not* absorbed.

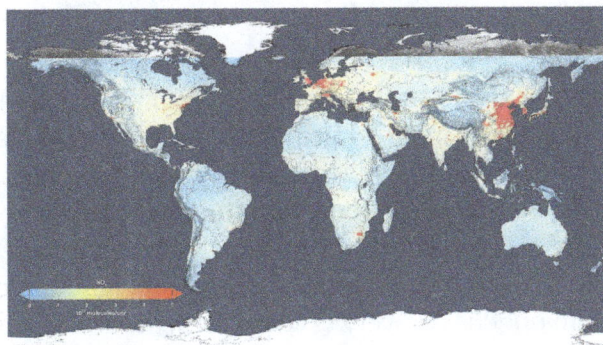

Nitrogen dioxide 2014 - global air quality levels.

Observations, lab measurements, and modeling are the three central elements in atmospheric chemistry. Progress in atmospheric chemistry is often driven by the interactions between these components and they form an integrated whole. For example, observations may tell us that more of a chemical compound exists than previously thought possible. This will stimulate new modelling and laboratory studies which will increase our scientific understanding to a point where the observations can be explained.

Observation

Observations of atmospheric chemistry are essential to our understanding. Routine observations of chemical composition tell us about changes in atmospheric composition over time. One important example of this is the Keeling Curve - a series of measurements from 1958 to today which show a steady rise in of the concentration of carbon dioxide. Observations of atmospheric chemistry are made in observatories such as that on Mauna Loa and on mobile platforms such as aircraft (e.g. the UK's Facility for Airborne Atmospheric Measurements), ships and balloons. Observations of atmospheric composition are increasingly made by satellites with important instruments such as GOME and MOPITT giving a global picture of air pollution and chemistry. Surface observations have the advantage that they provide long term records at high time resolution but are limited in the vertical and horizontal space they provide observations from. Some surface based instruments e.g. LIDAR can provide concentration profiles of chemical compounds and aerosol but are still restricted in the horizontal region they can cover. Many observations are available on line in Atmospheric Chemistry Observational Databases.

Laboratory Studies

Measurements made in the laboratory are essential to our understanding of the sources and sinks of pollutants and naturally occurring compounds. These experiments are performed in controlled environments that allow for the individual evaluation of specific chemical reactions or the assessment of properties of a particular atmospheric constituent. Types of analysis that are of interest includes both those on gas-phase reactions, as well as heterogeneous reactions that are relevant to the formation and growth of aerosols. Also of high importance is the study of atmospheric photochemistry which quantifies how the rate in which molecules are split apart by sunlight and what resulting products are. In addition, thermodynamic data such as Henry's law coefficients can also be obtained.

Modeling

In order to synthesise and test theoretical understanding of atmospheric chemistry, computer models (such as chemical transport models) are used. Numerical models solve the differential equations governing the concentrations of chemicals in the atmosphere. They can be very simple or very complicated. One common trade off in numerical models is between the number of chemical compounds and chemical reactions modelled versus the representation of transport and mixing in the atmosphere. For example, a box model might include hundreds or even thousands of chemical reactions but will only have a very crude representation of mixing in the atmosphere. In contrast, 3D models represent many of the physical processes of the atmosphere but due to constraints on computer resources will have far fewer chemical reactions and compounds. Models can be used to interpret observations, test understanding of chemical reactions and predict future concentrations of chemical compounds in the atmosphere. One important current trend is for atmospheric chemistry modules to become one part of earth system models in which the links between climate, atmospheric composition and the biosphere can be studied.

Some models are constructed by automatic code generators (e.g. Autochem or KPP). In this approach a set of constituents are chosen and the automatic code generator will then select the reactions involving those constituents from a set of reaction databases. Once the reactions have been chosen the ordinary differential equations (ODE) that describe their time evolution can be automatically constructed.

Null cycle

In atmospheric chemistry, a null cycle is a catalytic cycle that simply interconverts chemical species without leading to net production or removal of any component. In the stratosphere, null cycles are very important in contrasting series of reactions that lead to net depletion in the ozone layer.

One such cycle involves the nitrogen oxide species, which are the most responsible for ozone depletion in the stratosphere. The catalytic cycle is:

$$NO + O_3 \rightarrow NO_2 + O_2$$
$$NO_2 + O \rightarrow NO + O_2$$
$$\text{Net: } O + O_3 \rightarrow 2O_2$$

while the corresponding null cycle competes due to the possible photolysis of NO_2 which allows conservation of the odd oxygen species:

$$NO + O_3 \rightarrow NO_2 + O_2$$

$$NO_2 + hv \rightarrow NO + O$$

$$\text{Net: } O_3 + hv \rightarrow O + O_2$$

Since O and O_3 can exchange rapidly, the last cycle does not affect the rate of consumption of ozone which thus decreases during the day when photolysis can occur. NO can also react with other free radicals, such as chlorine and bromine, providing pathways that lead to null cycles:

$$Cl + O_3 \rightarrow ClO + O_2$$

$$ClO + NO \rightarrow Cl + NO_2$$

$$NO_2 + hv \rightarrow NO + O$$

$$\text{Net: } O_3 + hv \rightarrow O + O_2$$

Other null cycles, also termed *holding cycles*, produce reservoirs, effectively holding up the reactive species. An example is the formation of dinitrogen pentoxide:

$$NO_2 + NO_3 \rightarrow N_2O_5$$

This can lock up about 10% of the NO_x family of species present in the atmosphere, limiting their ability to participate in the ozone-destructing catalytic cycles.

Ozone–oxygen Cycle

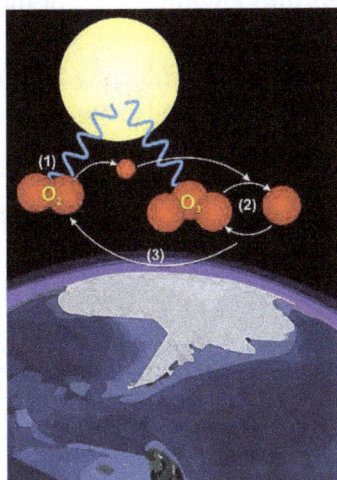

Ozone–oxygen cycle in the ozone layer: 1. Oxygen photolyzed to atomic oxygen 2. Oxygen and ozone continuously interconverted. Solar UV breaks down ozone; molecular and atomic oxygen combine. 3. Ozone is lost by reaction with atomic oxygen (plus other trace atoms).

The ozone–oxygen cycle is the process by which ozone is continually regenerated in Earth's stratosphere, converting ultraviolet radiation (UV) into heat. In 1930 Sydney Chapman resolved the chemistry involved. The process is commonly called the Chapman cycle by atmospheric scientists.

Most of the ozone production occurs in the tropical upper stratosphere and mesosphere. The total mass of ozone produced per day over the globe is about 400 million metric tons. The global mass of ozone is relatively constant at about 3 billion metric tons, meaning the Sun produces about 12% of the ozone layer each day.

Chemistry

1. Creation: an oxygen molecule is split (photolyzed) by higher frequency UV light (top end of UV-B, UV-C and above) into two oxygen atoms:

$$O_2 + hv \rightarrow 2\,O\bullet$$

Each oxygen atom then quickly combines with an oxygen molecule to form an ozone molecule:

$$O\bullet + O_2 \rightarrow O_3$$

2. The ozone–oxygen cycle: the ozone molecules formed by the reaction above absorb radiation having an appropriate wavelength between UV-C and UV-B. The triatomic ozone molecule becomes diatomic molecular oxygen plus a free oxygen atom:

$$O_3 + hv_{(240–310\ nm)} \rightarrow O_2 + O$$

The atomic oxygen produced quickly reacts with another oxygen molecule to reform ozone:

$$O + O_2 \rightarrow O_3 + E_K$$

where E_K denotes the excess energy of the reaction which is manifested as extra kinetic energy. These two reactions form the ozone–oxygen cycle, in which the chemical energy released when O and O_2 combine is converted into kinetic energy of molecular motion. The overall effect is to convert penetrating UV-B light into heat, without any net loss of ozone. This cycle keeps the ozone layer in a stable balance while protecting the lower atmosphere from UV radiation, which is harmful to most living beings. It is also one of two major sources of heat in the stratosphere (the other being the kinetic energy released when O_2 is photolyzed into O atoms).

3. Removal: if an oxygen atom and an ozone molecule meet, they recombine to form two oxygen molecules:

$$O_3 + O\bullet \rightarrow 2\,O_2$$

And if two oxygen atoms meet, they react to form one oxygen molecule:

$$2\,O\bullet \rightarrow O_2$$

This reaction is known to have a negative order of reaction of -1. The overall amount of ozone in the stratosphere is determined by a balance between production by solar radiation and removal. The removal rate is slow, since the concentration of O atoms is very low.

1. The net reaction will be $2\,O_3 \rightarrow 3\,O_2$

Certain free radicals, the most important being hydroxyl (OH), nitric oxide (NO) and atoms of chlorine (Cl) and bromine (Br), catalyze the recombination reaction, leading to an ozone layer that is thinner than it would be if the catalysts were not present.

Most of the OH and NO are naturally present in the stratosphere, but human activity, especially emissions of chlorofluorocarbons (CFCs) and halons, has greatly increased the Cl and Br concentrations, leading to ozone depletion. Each Cl or Br atom can catalyze tens of thousands of decomposition reactions before it is removed from the stratosphere.

Leighton Relationship

In atmospheric chemistry, the Leighton relationship is an equation that determines the concentration of tropospheric ozone in areas polluted by the presence of nitrogen oxides. Ozone in the troposphere is primarily produced through the photolysis of nitrogen dioxide at wavelengths (λ) less than 430 nm, which are able to reach the lowest levels of the atmosphere, through the following mechanism:

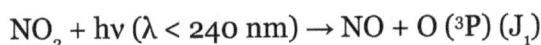

$$NO_2 + h\nu\ (\lambda < 240\ nm) \rightarrow NO + O\ (^3P)\ (J_1)$$

$$O\ (^3P) + O_2 + M \rightarrow O_3 + M\ (k_2)$$

$$NO + O_3 \rightarrow NO_2 + O_2\ (k_3)$$

Since O (^{3}P) is very reactive it can be assumed to be in steady state, and thus an equation linking the concentrations of the species involved can be derived:

$$[O_3] = J_1[NO_2]/k_3[NO]$$

The Leighton relationship above shows how production of ozone is directly related to the solar intensity and hence to the zenith angle. The yield of this molecule will therefore be a maximum during the day, especially at noon and in the summer season; it also demonstrates how high concentrations of both ozone and nitric oxide are unfeasible. However, NO can react with peroxyl radicals to give back NO_2 without loss of ozone:

$$RO_2 + NO \rightarrow NO_2 + RO$$

providing another pathway to allow the buildup of O_3.

This relationship is named after Philip Leighton, who wrote a significant book in 1961 describing air pollution, as recognition of his contributions in the understanding of tropospheric chemistry.

Dobson Unit

The Dobson unit (DU) is a unit of measurement of the amount of a trace gas in a vertical column through the Earth's atmosphere. It originated, and continues to be primarily used in respect to, atmospheric ozone, whose total column amount, usually termed "total ozone", and sometimes "column abundance", is dominated by the high concentrations of ozone in the stratospheric ozone layer. One Dobson unit is equal to the number of ozone molecules needed to create a pure layer of ozone 0.01 millimeters thick at STP - standard conditions for temperature and pressure.

The Dobson Unit is defined as the thickness (in units of 10 µm) of that layer of pure gas which would be formed by the total column amount at STP. This is sometimes referred to as a 'milli-at-mo-centimeter.' A typical column amount of 300 DU of atmospheric ozone therefore would form a 3 mm layer of pure gas at the surface of the Earth if its temperature and pressure conformed to STP. The Dobson unit is named after Gordon Dobson, a researcher at the University of Oxford who in the 1920s built the first instrument to measure total ozone from the ground, making use of a double prism monochromator to measure the differential absorption of different bands of solar ultraviolet radiation by the ozone layer. This instrument, called the Dobson ozone spectrophotometer, has formed the backbone of the global network for monitoring atmospheric ozone and was the source of the discovery in 1984 of the Antarctic ozone hole.

Relation to SI Units

The Dobson Unit is not part of the SI International System of Units. To address this shortcoming, a brief study in 1982 examined a number of alternative SI-based units suitable for column amounts of not only ozone but any gas in any planetary atmosphere and proposed the use of the unit of mole per square metre for all cases. Examples range from Earth's trace gases at levels of micro moles per square meter to Venus's carbon dioxide at mega moles per square meter. Typical values of total ozone in the Earth's atmosphere are conveniently represented in millimoles per square metre (mmol m-2.) One DU is equivalent to 0.4462 mmol m-2. One DU is also equivalent to 2.687×10^{20} molecules per square metre.

A later examination in 1995 of units for use in atmospheric chemistry by the Commission on Atmospheric Chemistry, a part of the International Union of Pure and Applied Chemistry, discouraged the use of special names and symbols for units that are not part of the SI and are not products of powers of SI base units. Although it overlooked the 1982 article it concurred with the view that the Dobson unit should eventually be replaced by an appropriate SI unit and that the unit mmol m-2 was the most convenient and least cumbersome option. It expressed the hope that this unit would eventually supplant the Dobson Unit. However, as of March 2017, there is little evidence that this has occurred; for example the Dobson unit is still used by NASA and by the World Ozone and Ultraviolet Radiation Data Center in their reporting of total ozone amount.

Ozone

NASA uses a baseline value of 220 DU for ozone. This was chosen as the starting point for observations of the Antarctic ozone hole, since values of less than 220 Dobson units were not found before 1979. Also, from direct measurements over Antarctica, a column ozone level of less than 220 Dobson units is a result of the ozone loss from chlorine and bromine compounds.

Sulfur Dioxide

In addition, Dobson units are often used to describe total column densities of sulfur dioxide, which occurs in the atmosphere in small amounts due to the combustion of fossil fuels, from biological processes releasing dimethyl sulfide, or by natural combustion such as forest fires. Large amounts of sulfur dioxide may be released into the atmosphere as well by volcanic eruptions. The Dobson unit is used to describe total column amounts of sulfur dioxide because it appeared in the early days of ozone remote sensing on ultraviolet satellite instruments (such as TOMS).

Derivation

The Dobson Unit arises from the ideal gas law. From the real gas law:

$$PV = nRT$$

where P and V are pressure and volume, respectively, and n, R and T are the number of moles of gas, the gas constant (8.314 J/mol K), and T is temperature in Kelvin (K).

The number density of air is the number of molecules or atoms per unit volume:

$$n_{air} = \frac{A_{av}n}{V}$$

and when plugged into the real gas law, the number density of air is found by using pressure, temperature and the real gas constant.

$$n_{air} = \frac{A_{av}P}{RT}$$

The number density (molecules/volume) of air at standard temperature and pressure (T = 273K and P = 101325 Pa) is, by using this equation:

$$n_{air} = \frac{A_{av}P}{RT} = \frac{(6.02 \times 10^{23} \frac{molecules}{mol}) \cdot (101325 Pa)}{8.314 \frac{J}{molK} \cdot 273K}$$

With some unit conversions of Joules to Pascals, the equation for molecules/volume is

$$\frac{(6.02 \times 10^{23} \frac{molecules}{mol}) \cdot (101325 Pa)}{8.314 \frac{Pa\,m^3}{molK} \cdot 273K} = 2.69 \times 10^{25}\,molecules\,m^{-3}$$

A Dobson Unit is the total amount of a trace gas per unit area. In atmospheric sciences, this is referred to as a column density. How, though, do we go from units of molecules per cubic meter, a volume, to molecules per square centimeter, an area? This must be done by integration. To get a column density, we must integrate the total column over a height. Per the definition of Dobson Units, we see that 1 DU = 0.01 mm of trace gas when compressed down to sea level at standard temperature and pressure. So if we integrate our number density of air from 0 to 0.01 mm, we find the number density which is equal to 1 DU:

$$\int_{0mm}^{0.01mm} (2.69 \times 10^{25}\,molecules\,m^{-3})dx = 2.69 \times 10^{25}\,molecules\,m^{-3} \cdot 0.01mm - 2.69 \times 10^{25}\,molecules\,m^{-3} \cdot 0mm$$

$$= 2.69 \times 10^{25}\,molecules\,m^{-3} \cdot 10^{-5}\,m = 2.69 \times 10^{20}\,molecules\,m^{-2}$$

And thus we come up with the value of 1 DU, which is 2.69×10[20] molecules per meter squared.

Climate Oscillation

A climate oscillation or climate cycle is any recurring cyclical oscillation within global or regional climate, and is a type of climate pattern. These fluctuations in atmospheric temperature, sea surface temperature, precipitation or other parameters can be quasi-periodic, often occurring on inter-annual, multi-annual, decadal, multidecadal, century-wide, millennial or longer timescales. They are not perfectly periodic and a Fourier analysis of the data does not give a sharp spectrum.

A prominent example is the El Niño Southern Oscillation, involving sea surface temperatures along a stretch of the equatorial Central and East Pacific Ocean and the western coast of tropical South America, but which affects climate worldwide.

Records of past climate conditions are recovered through geological examination of proxies, found in glacier ice, sea bed sediment, tree ring studies or otherwise.

Examples

Many oscillations on different time-scales are hypothesized, although the causes may be unknown. (Some of them are more like a random walk than an oscillation.) Here is a list of known or proposed climatic oscillations:

- the glacial periods of the last ice age – period around 100 000 years (Quaternary glaciation#Astronomical cycles and 100,000-year problem)
- North African climate cycles – tens of thousands of years
- the Atlantic Multidecadal Oscillation – around 50 to 70 years, but unpredictable
- the El Niño Southern Oscillation – 2 to 7 years
- the Pacific decadal oscillation – 8 to 12 years? (not clear)
- the Interdecadal Pacific Oscillation – 15 to 30 years? (not clear)
- the Arctic oscillation – no particular periodicity
- the North Atlantic Oscillation – no particular periodicity
- the North Pacific Oscillation – ?
- the Hale cycle or sunspot cycle – about 11 years (may be discernible in climate records; solar variation)
- the Quasi-biennial oscillation – about 30 months
- a 60-year climate cycle recorded in many ancient calendars

Anomalies in oscillations sometimes occur when they coincide, as in the Arctic dipole anomaly (a combination of the Arctic and North Atlantic oscillations) and the longer-term Younger Dryas, a sudden non-linear cooling event that occurred at the onset of the current Holocene interglacial. In the case of volcanoes, large eruptions such as Mount Tambora in 1816, which led to the Year Without a Summer, typically cool the climate, especially when the volcano is located in the tropics.

Around 70 000 years ago the Toba supervolcano eruption created an especially cold period during the ice age, leading to a possible genetic bottleneck in human populations. However, outgassing from large igneous provinces such as the Permian Siberian Traps can input carbon dioxide into the atmosphere, warming the climate. Triggering of other mechanisms, such as methane clathrate deposits as during the Paleocene-Eocene Thermal Maximum, increased the rate of climatic temperature change and oceanic extinctions.

Another longer-term near-millennial oscillation involves the Daansgard-Oeschger cycles, occurring on roughly 1,500-year cycles during the last glacial maximum. They may be related to the Holocene Bond events, and may involve factors similar to those responsible for Heinrich events.

Origins and Causes

There are close correlations between Earth's climate oscillations and astronomical factors (barycenter changes, solar variation, cosmic ray flux, cloud albedo feedback, Milankovic cycles), and modes of heat distribution between the ocean-atmosphere climate system. In some cases, current, historical and paleoclimatological natural oscillations may be masked by significant volcanic eruptions, impact events, irregularities in climate proxy data, positive feedback processes or anthropogenic emissions of substances such as greenhouse gases.

Effects

Extreme phases of short-term climate oscillations such as ENSO can result in characteristic patterns of floods and droughts (including megadroughts), monsoonal disruption and extreme temperatures in the form of heat waves and cold waves. Shorter-term climate oscillations typically do not directly result in longer-term climate change in temperatures. However, the effects of underlying climate trends such as recent global warming and oscillations can be cumulative to global temperature, producing shorter-term fluctuations in the instrumental and satellite temperature records.

Collapses of past civilizations such as the Maya may be related to cycles of precipitation, especially drought, that in this example also correlates to the Western Hemisphere Warm Pool.

One example of possible correlations between factors affecting the climate and global events, popular with the media, is a 2003 study on the correlation between wheat prices and sunspot numbers.

Analysis and Uncertainties

Radiative forcings and other factors in a climate oscillation must obey the laws of atmospheric thermodynamics. However, because Earth's climate is inherently a complex system, simple Fourier analysis or climate modelling often does not create a perfect replication of the observed or inferred conditions. No climate cycle is found to be perfectly periodic, although the Milankovich cycles (based on multiple superimposed orbital cycles and Earth's precession) are quite close to being periodic (perhaps almost periodic?).

One difficulty in detecting climate cycles is that the Earth's climate has been changing in non-cyclic ways over most paleoclimatological timescales. For instance, we are now in a period of global warming that appears anthropogenic. In a larger timeframe, the Earth is emerging from the latest ice age, cooling from the Holocene climatic optimum and warming from the so-called "Little Ice Age", which means that climate has been constantly changing over the last 15,000 years or so. During warm periods, temperature fluctuations are often of a lesser amplitude. The Pleistocene period, dominated by repeated glaciations, developed out of more stable conditions in the Miocene and Pliocene climate. Holocene climate has been relatively stable. All of these changes complicate the task of looking for cyclical behavior in the climate.

Positive feedback, negative feedback, and ecological inertia from the land-ocean-atmosphere system often attenuate or reverse smaller effects, whether from orbital forcings, solar variations or changes in concentrations of greenhouse gases. Most climatologists recognize the existence of various tipping points that push small forcings beyond a certain threshold that makes the change irreversible while the forcings are still in place. Certain feedbacks involvong processes such as clouds are also uncertain; for contrails, natural cirrus clouds, oceanic dimethyl sulfide and a land-based equivalent, competing theories exist concerning effects on climatic temperatures, for example contrasting the Iris hypothesis and CLAW hypothesis.

Through Geologic and Historical Time

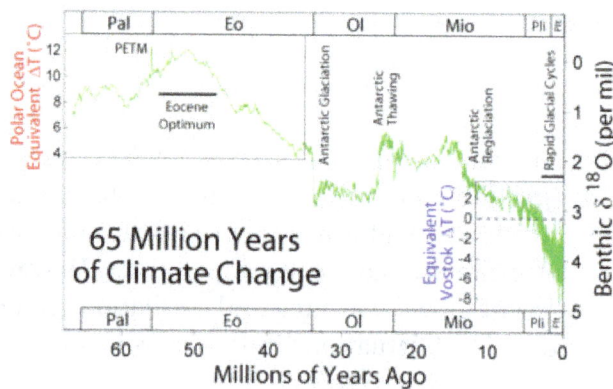

Climate change over the past 65 million years, using proxy data including Oxygen-18 ratios from foraminifera.

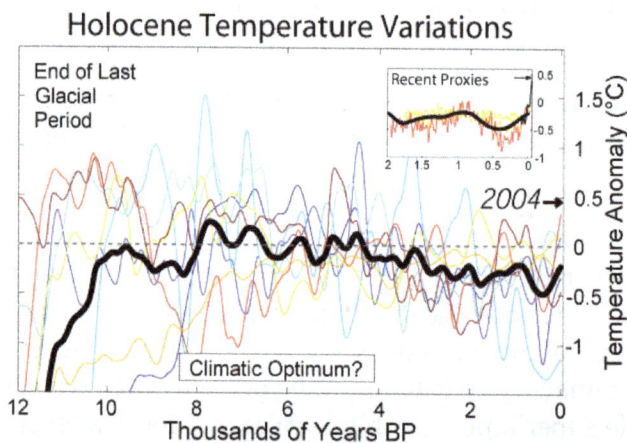

Temperature change over the past 12 000 years, from various sources. The thick black curve is an average.

Various climate forcings are typically in flux throughout geologic time, and some processes of the Earth's temperature may be self-regulating. For example, during the Snowball Earth period, large glacial ice sheets spanned to Earth's equator, covering nearly its entire surface, and very low albedo created extremely low temperatures, while the accumulation of snow and ice likely removed carbon dioxide through atmospheric deposition. However, the absence of plant cover to absorb atmospheric CO_2 emitted by volcanoes meant that the greenhouse gas could accumulate in the atmosphere. There was also an absence of exposed silicate rocks, which use CO_2 when they undergo weathering. This created a warming that later melted the ice and brought Earth's temperature back to equilibrium. During the following eons of the Paleozoic, cosmic ray flux and occasional nearby supernova explosions (one hypothesis for the cause of the Ordovician–Silurian extinction event) and gamma ray bursts may have induced ice ages or other sudden climate changes.

Throughout the Cenozoic, multiple climate forcings led to warming and cooling of the atmosphere, which led to the early formation of the Antarctic ice sheet, subsequent melting, and its later reglaciation. The temperature changes occurred somewhat suddenly, at carbon dioxide concentrations of about 600–760 ppm and temperatures approximately 4°C warmer than today. During the Pleistocene, cycles of glaciations and interglacials occurred on cycles of roughly 100,000 years, but may stay longer within an interglacial when orbital eccentricity approaches zero, as during the current interglacial. Previous interglacials such as the Eemian phase created temperatures higher than today, higher sea levels, and some partial melting of the West Antarctic ice sheet. The warmest part of the current interglacial occurred during the early Holocene Optimum, when temperatures were a few degrees Celsius warmer than today, and a strong African Monsoon created grassland conditions in the Sahara during the Neolithic Subpluvial. Since that time, several cooling events have occurred, including:

- the Piora Oscillation
- the Middle Bronze Age Cold Epoch
- the Iron Age Cold Epoch
- cooling during the Dark Ages
- the Spörer Minimum
- the "Little Ice Age"
- the Dalton Minimum
- volcanic coolings such as from Laki in Iceland
- the phase of cooling c. 1940-1970, which led to global cooling hypotheses

In contrast, several warm periods have also taken place, and they include but are not limited to:

- the Older Peron during the late Holocene optimum
- a warm period during the apex of the Minoan civilization
- the Roman Warm Period
- the Medieval Warm Period
- the retreat of glaciers since 1850
- the "Modern Warming" during the 20th century

Certain effects have occurred during these cycles. For example, during the Medieval Warm Period, the American Midwest was in drought, including the Sand Hills of Nebraska which were active sand dunes. The black death plague of *Yersinia pestis* also occurred during Medieval temperature fluctuations, and may be related to changing climates.

Given that records of solar activity are accurate, solar activity may have contributed to part of the modern warming that peaked in the 1930s, in addition to the 60-year temperature cycles that result in roughly 0.5°C of warming during the increasing temperature phase. However, solar cycles fail to account for warming observed since the 1980s to the present day. Events such as the opening of the Northwest Passage and recent record low ice minima of the modern Arctic shrinkage have not taken place for at least several centuries, as early explorers were all unable to make an Arctic crossing, even in summer. Shifts in biomes and habitat ranges are also unprecedented, occurring at rates that do not coincide with known climate oscillations. The extinction of many tropical amphibian species, especially in cloud forests, have been attributed to changing global temperatures, fungal disease and possible influence from unusually extreme phases of oceanic climate oscillations.

Climate Pattern

A climate pattern is any recurring characteristic of the climate. Climate patterns can last tens of thousands of years, like the glacial and interglacial periods within ice ages, or repeat each year, like monsoons.

A climate pattern may come in the form of a regular cycle, like the diurnal cycle or the seasonal cycle; a quasi periodic event, like El Niño; or a highly irregular event, such as a volcanic winter. The regular cycles are generally well understood and may be removed by normalization. For example, graphs which show trends of temperature change will usually have the effects of seasonal variation removed.

Modes of Variability

A *mode of variability* is a climate pattern with identifiable characteristics, specific regional effects, and often oscillatory behavior. Many modes of variability are used by climatologists as indices to represent the general climatic state of a region affected by a given climate pattern.

Measured via an empirical orthogonal function analysis, the mode of variability with the greatest effect on climates worldwide is the seasonal cycle, followed by El Niño-Southern Oscillation, followed by thermohaline circulation.

Other well-known modes of variability include:

- The Antarctic oscillation
- The Arctic oscillation
- The Atlantic multidecadal oscillation
- The Indian Ocean Dipole
- The Madden–Julian oscillation
- The North Atlantic oscillation

- The Pacific decadal oscillation

- The Pacific-North American teleconnection pattern

- The Quasi-biennial oscillation

Radiative Forcing

Radiative forcing or climate forcing is the difference between insolation (sunlight) absorbed by the Earth and energy radiated back to space. Typically, radiative forcing is quantified at the tropopause in units of watts per square meter of the Earth's surface. Positive forcing (incoming energy exceeding outgoing energy) warms the system, while negative forcing (outgoing energy exceeding incoming energy) cools it. Causes of radiative forcing include changes in insolation and the concentrations of radiatively active gases, commonly known as greenhouse gases and aerosols.

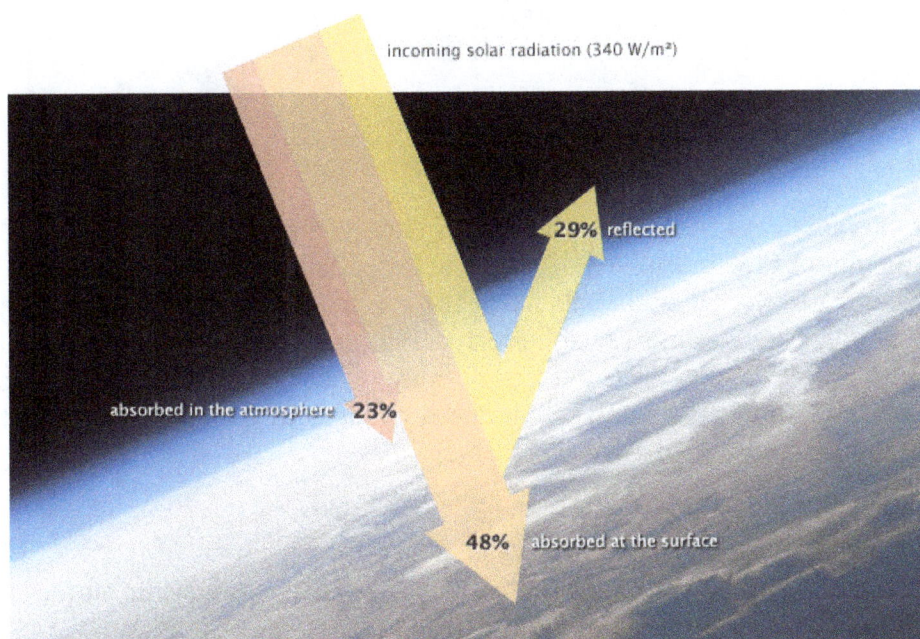

Incoming solar radiation

Radiation Balance

Atmospheric gases only absorb some wavelengths of energy but are transparent to others. The absorption patterns of water vapor (blue peaks) and carbon dioxide (pink peaks) overlap in some wavelengths. Carbon dioxide is not as strong a greenhouse gas as water vapor, but it absorbs energy in wavelengths (12-15 micrometers) that water vapor does not, partially closing the "window" through which heat radiated by the surface would normally escape to space.

Almost all of the energy that affects Earth's climate is received as radiant energy from the Sun. The planet and its atmosphere absorb and reflect some of the energy, while long-wave energy is radiated back into space. The balance between absorbed and radiated energy determines the average global temperature. Because the atmosphere absorbs some of the re-radiated long-wave energy, the planet is warmer than it would be in the absence of the atmosphere which is greenhouse effect.

IPCC usage

The Intergovernmental Panel on Climate Change (IPCC) AR4 report defines radiative forcings as:

"Radiative forcing is a measure of the influence a factor has in altering the balance of incoming and outgoing energy in the Earth-atmosphere system and is an index of the importance of the factor as a potential climate change mechanism. In this report radiative forcing values are for changes relative to preindustrial conditions defined at 1750 and are expressed in Watts per square meter (W/m²)."

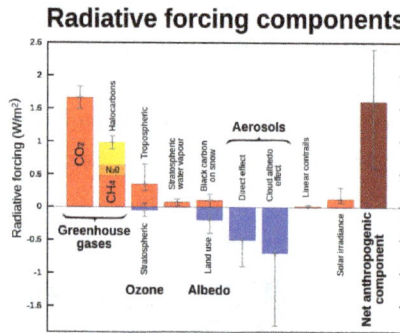

Radiative forcings, IPCC 2007.

In simple terms, radiative forcing is "...the rate of energy change per unit area of the globe as measured at the top of the atmosphere." In the context of climate change, the term "forcing" is restricted to changes in the radiation balance of the surface-troposphere system imposed by external factors, with no changes in stratospheric dynamics, no surface and tropospheric feedbacks in operation (i.e., no secondary effects induced because of changes in tropospheric motions or its thermodynamic state), and no dynamically induced changes in the amount and distribution of atmospheric water (vapour, liquid, and solid forms).

Climate Sensitivity

Radiative forcing can be used to estimate a subsequent change in equilibrium surface temperature (ΔT_s) arising from that forcing via the equation:

$$\Delta T_s = \lambda \, \Delta F$$

where λ is the climate sensitivity, usually with units in K/(W/m²), and ΔF is the radiative forcing. A typical value of λ is 0.8 K/(W/m²), which gives a warming of 3K for doubling of CO_2.

Example Calculations

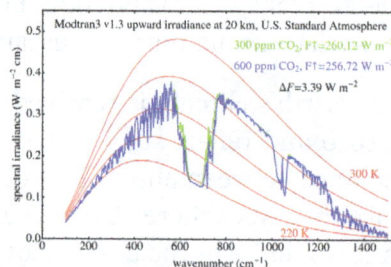

Radiative forcing for doubling CO_2, as calculated by radiative transfer code Modtran. Red lines are Planck curves.

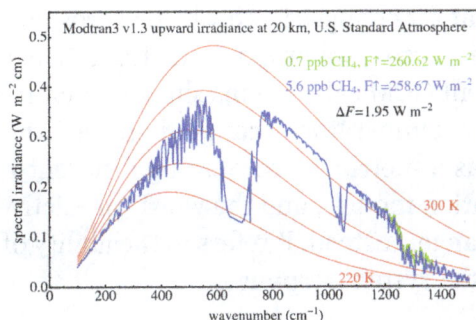

Radiative forcing for eight times increase of CH_4, as calculated by radiative transfer code Modtran.

Solar Forcing

Radiative forcing (measured in Watts per square meter) can be estimated in different ways for different components. For solar irradiance (*i.e.*, "solar forcing"), the radiative forcing is simply the change in the average amount of solar energy absorbed per square meter of the Earth's area. Since the Earth's cross-sectional area exposed to the Sun (πr^2) is equal to 1/4 of the surface area of the Earth ($4\pi r^2$), the solar input per unit area is one quarter the change in solar intensity. This must be multiplied by the fraction of incident sunlight that is absorbed, F=(1-R), where R is the reflectivity (albedo), of the Earth. The albedo is approximately 0.3, so F is approximately equal to 0.7. Thus, the solar forcing is the change in the solar intensity divided by 4 and multiplied by 0.7.

Likewise, a change in albedo will produce a solar forcing equal to the change in albedo divided by 4 multiplied by the solar constant.

Forcing Due to Atmospheric Gas

For a greenhouse gas, such as carbon dioxide, radiative transfer codes that examine each spectral line for atmospheric conditions can be used to calculate the change ΔF as a function of changing concentration. These calculations can often be simplified into an algebraic formulation that is specific to that gas.

For instance, the simplified first-order approximation expression for carbon dioxide is:

$$\ddot{A}F = 5.35 \times \ln \frac{C}{C_0} \text{ W m}^{-2}$$

where C is the CO_2 concentration in parts per million by volume and C_0 is the reference concentration. The relationship between carbon dioxide and radiative forcing is logarithmic and thus increased concentrations have a progressively smaller warming effect.

A different formula applies for other greenhouse gases such as methane and N_2O (square-root dependence) or CFCs (linear), with coefficients that can be found *e.g.* in the IPCC reports.

Related Measures

Radiative forcing is a useful way to compare different causes of perturbations in a climate system. Other possible tools can be constructed for the same purpose: for example Shine *et al.* say "...

recent experiments indicate that for changes in absorbing aerosols and ozone, the predictive ability of radiative forcing is much worse... we propose an alternative, the 'adjusted troposphere and stratosphere forcing'. We present GCM calculations showing that it is a significantly more reliable predictor of this GCM's surface temperature change than radiative forcing. It is a candidate to supplement radiative forcing as a metric for comparing different mechanisms...". In this quote, GCM stands for "global circulation model", and the word "predictive" does not refer to the ability of GCMs to forecast climate change. Instead, it refers to the ability of the alternative tool proposed by the authors to help explain the system response.

History

The table below (derived from atmospheric radiative transfer models) shows changes in radiative forcing between 1979 and 2013. The table includes the contribution to radiative forcing from carbon dioxide (CO_2), methane (CH_4), nitrous oxide (N_2O); chlorofluorocarbons (CFCs) 12 and 11; and fifteen other minor, long-lived, halogenated gases. The table includes the contribution to radiative forcing of long-lived greenhouse gases. It does not include other forcings, such as aerosols and changes in solar activity.

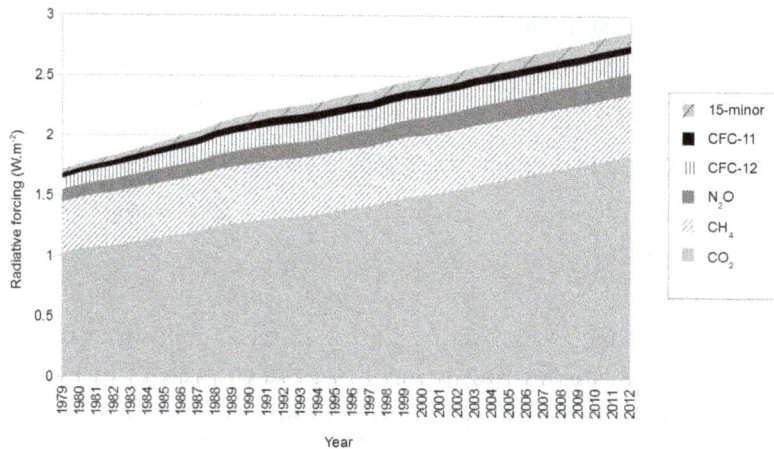

Changes in radiative forcing of long-lived greenhouse gases between 1979 and 2012.

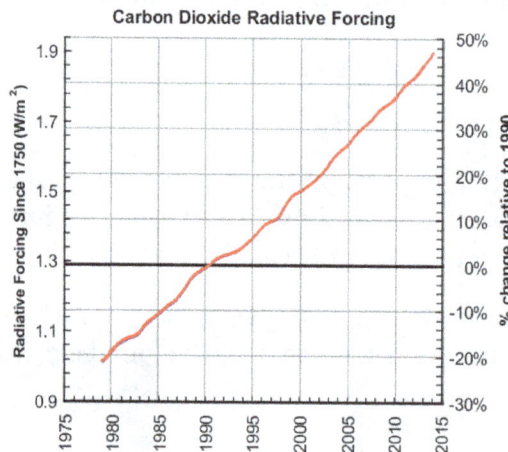

Radiative forcing, relative to 1750, due to carbon dioxide alone since 1979. The percent change
from January 1, 1990 is shown on the right axis.

Global radiative forcing (relative to 1750, in $W\,m^{-2}$), CO_2-equivalent mixing ratio, and the Annual Greenhouse Gas Index (AGGI) between 1979–2014									
Year	CO_2	CH_4	N_2O	CFC-12	CFC-11	15-minor	Total	CO_2-eq ppm	AGGI 1990 = 1	AGGI % change
1979	1.027	0.419	0.104	0.092	0.039	0.031	1.712	383	0.786	
1980	1.058	0.426	0.104	0.097	0.042	0.034	1.761	386	0.808	2.8
1981	1.077	0.433	0.107	0.102	0.044	0.036	1.799	389	0.826	2.2
1982	1.089	0.440	0.111	0.108	0.046	0.038	1.831	391	0.841	1.8
1983	1.115	0.443	0.113	0.113	0.048	0.041	1.873	395	0.860	2.2
1984	1.140	0.446	0.116	0.118	0.050	0.044	1.913	397	0.878	2.2
1985	1.162	0.451	0.118	0.123	0.053	0.047	1.953	401	0.897	2.1
1986	1.184	0.456	0.122	0.129	0.056	0.049	1.996	404	0.916	2.2
1987	1.211	0.460	0.120	0.135	0.059	0.053	2.039	407	0.936	2.2
1988	1.250	0.464	0.123	0.143	0.062	0.057	2.099	412	0.964	3.0
1989	1.274	0.468	0.126	0.149	0.064	0.061	2.144	415	0.984	2.1
1990	1.293	0.472	0.129	0.154	0.065	0.065	2.178	418	1.000	1.6
1991	1.313	0.476	0.131	0.158	0.067	0.069	2.213	420	1.016	1.6
1992	1.324	0.480	0.133	0.162	0.067	0.072	2.238	422	1.027	1.1
1993	1.334	0.481	0.134	0.164	0.068	0.074	2.254	424	1.035	0.7
1994	1.356	0.483	0.134	0.166	0.068	0.075	2.282	426	1.048	1.3
1995	1.383	0.485	0.136	0.168	0.067	0.077	2.317	429	1.064	1.5
1996	1.410	0.486	0.139	0.169	0.067	0.078	2.350	431	1.079	1.4
1997	1.426	0.487	0.142	0.171	0.067	0.079	2.372	433	1.089	0.9
1998	1.465	0.491	0.145	0.172	0.067	0.080	2.419	437	1.111	2.0
1999	1.495	0.494	0.148	0.173	0.066	0.082	2.458	440	1.128	1.6
2000	1.513	0.494	0.151	0.173	0.066	0.083	2.481	442	1.139	0.9
2001	1.535	0.494	0.153	0.174	0.065	0.085	2.506	444	1.150	1.0
2002	1.564	0.494	0.156	0.174	0.065	0.087	2.539	447	1.166	1.3
2003	1.601	0.496	0.158	0.174	0.064	0.088	2.580	450	1.185	1.6
2004	1.627	0.496	0.160	0.174	0.063	0.090	2.610	453	1.198	1.1
2005	1.655	0.495	0.162	0.173	0.063	0.092	2.640	455	1.212	1.2
2006	1.685	0.495	0.165	0.173	0.062	0.095	2.675	458	1.228	1.3
2007	1.710	0.498	0.167	0.172	0.062	0.097	2.706	461	1.242	1.1
2008	1.739	0.500	0.170	0.171	0.061	0.100	2.742	464	1.259	1.3
2009	1.760	0.502	0.172	0.171	0.061	0.103	2.768	466	1.271	1.0
2010	1.791	0.504	0.174	0.170	0.060	0.106	2.805	470	1.288	1.3
2011	1.818	0.505	0.178	0.169	0.060	0.109	2.838	473	1.303	1.2
2012	1.846	0.507	0.181	0.168	0.059	0.111	2.873	476	1.319	1.2
2013	1.884	0.509	0.184	0.167	0.059	0.114	2.916	479	1.338	1.5
2014	1.909	0.500	0.187	0.166	0.058	0.116	2.936	481	1.356	1.6

The table shows that CO_2 dominates the total forcing, with methane and chlorofluorocarbons (CFC) becoming relatively smaller contributors to the total forcing over time. The five major greenhouse gases account for about 96% of the direct radiative forcing by long-lived greenhouse gas increases since 1750. The remaining 4% is contributed by the 15 minor halogenated gases.

The table also includes an "Annual Greenhouse Gas Index" (AGGI), which is defined as the ratio of the total direct radiative forcing due to long-lived greenhouse gases for any year for which adequate global measurements exist to that which was present in 1990. 1990 was chosen because it is the baseline year for the Kyoto Protocol. This index is a measure of the inter-annual changes in conditions that affect carbon dioxide emission and uptake, methane and nitrous oxide sources and sinks, the decline in the atmospheric abundance of ozone-depleting chemicals related to the Montreal Protocol. and the increase in their substitutes (hydrogenated CFCs (HCFCs) and hydrofluorocarbons (HFC). Most of this increase is related to CO_2. For 2013, the AGGI was 1.34 (representing an increase in total direct radiative forcing of 34% since 1990). The increase in CO_2 forcing alone since 1990 was about 46%. The decline in CFCs considerably tempered the increase in net radiative forcing.

References

- Farman, J. C.; Gardiner, B. G.; Shankin, J. D. (1985). "Large losses of total ozone in Antarctica reveal seasonal ClO_x/NO_x interaction". Nature. 315 (16 May 1985): 207–210. doi:10.1038/315207a0

- National Academies of Sciences, Engineering, and Medicine (2016). Future of Atmospheric Research: Remembering Yesterday, Understanding Today, Anticipating Tomorrow. Washington, DC: The National Academies Press. p. 15. ISBN 978-0-309-44565-8

- St. Fleur, Nicholas (10 November 2015). "Atmospheric Greenhouse Gas Levels Hit Record, Report Says". New York Times. Retrieved 11 November 2015

- Scafetta, Nicola (May 15, 2010). "Empirical evidence for a celestial origin of the climate oscillations" (PDF). Journal of Atmospheric and Solar-Terrestrial Physics. 72: 951–970. Bibcode:2010JASTP..72..951S. arXiv:1005.4639

- S. E. Schwartz; P. Warneck (1995). "Units for use in atmospheric chemistry". Pure Appl. Chem. 67 (8-9): 1377–1406. doi:10.1351/pac199567081377

- John Roger Barker (1995). Progress And Problems In Atmospheric Chemistry. World Scientific. p. 22. ISBN 9789810221133

- Cole, Steve; Gray, Ellen (14 December 2015). "New NASA Satellite Maps Show Human Fingerprint on Global Air Quality". NASA. Retrieved 14 December 2015

- Shine et al., An alternative to radiative forcing for estimating the relative importance of climate change mechanisms, Geophysical Research Letters, Vol 30, No. 20, 2047, doi:10.1029/2003GL018141, 2003

- James Pfafflin; Edward Ziegler (2006). Encyclopedia of Environmental Science And Engineering. 1. CRC Press. p. 122. ISBN 9780849398438

- Zimmer, Carl (3 October 2013). "Earth's Oxygen: A Mystery Easy to Take for Granted". New York Times. Retrieved 3 October 2013

4

Physical Properties of the Earth's Atmosphere

The physical properties of the Earth's atmosphere discussed in the section are atmospheric pressure, atmospheric temperature, sunlight and density of air. Atmospheric pressure is the pressure at any given point in the Earth's atmosphere whereas atmospheric temperature is the temperature of the atmosphere decided by phenomena like humidity and altitude. The chapter serves as a source to understand the major categories related to the atmosphere of Earth.

Atmospheric Pressure

Atmospheric pressure, sometimes also called barometric pressure, is the pressure within the atmosphere of Earth (or that of another planet). In most circumstances atmospheric pressure is closely approximated by the hydrostatic pressure caused by the weight of air above the measurement point. As elevation increases, there is less overlying atmospheric mass, so that atmospheric pressure decreases with increasing elevation. Pressure measures force per unit area, with SI units of pascals (1 Pa = 1 N/m²). On average, a column of air one square centimetre [cm²] (0.16 sq in) in cross-section, measured from sea level to the top of the Earth's atmosphere, has a mass of about 1.03 kilograms (2.3 lb) and weight of about 10.1 newtons (2.3 lb$_f$). That weight (across one square centimeter) is a pressure of 10.1 N/cm² or 101 kN/m² (kPa). A column 1 square inch (6.5 cm²) in cross-section would have a weight of about 14.7 lb (6.7 kg) or about 65.4 N.

Mechanism

Atmospheric pressure is caused by the gravitational attraction of the planet on the atmospheric gases above the surface, and is a function of the mass of the planet, the radius of the surface, and the amount of gas and its vertical distribution in the atmosphere. It is modified by the planetary rotation and local effects such as wind velocity, density variations due to temperature and variations in composition.

Standard Atmosphere

The standard atmosphere (symbol: atm) is a unit of pressure defined as 101325 Pa (1.01325 bar), equivalent to 760 mmHg (torr), 29.92 inHg and 14.696 psi.

Mean Sea Level Pressure

The mean sea level pressure (MSLP) is the average atmospheric pressure at sea level. This is the atmospheric pressure normally given in weather reports on radio, television, and newspapers or on the Internet. When barometers in the home are set to match the local weather reports, they measure pressure adjusted to sea level, not the actual local atmospheric pressure.

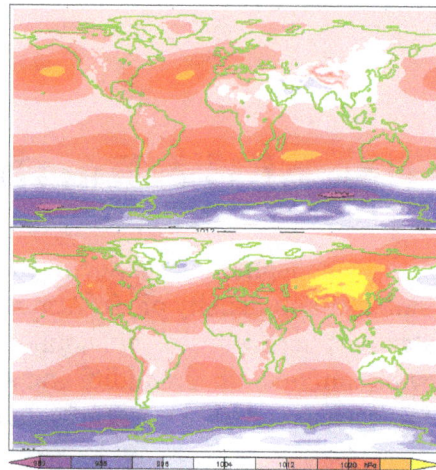

15 year average mean sea level pressure for June, July, and August (top) and December, January, and February (bottom). ERA-15 re-analysis.

Kollsman-type barometric aircraft altimeter (as used in North America) displaying an altitude of 80 ft (24 m).

The *altimeter setting* in aviation, is an atmospheric pressure adjustment.

Average *sea-level pressure* is 1013.25 mbar (101.325 kPa; 29.921 inHg; 760.00 mmHg). In aviation weather reports (METAR), QNH is transmitted around the world in millibars or hectopascals (1 hectopascal = 1 millibar), except in the United States, Canada, and Colombia where it is reported in inches (to two decimal places) of mercury. The United States and Canada also report *sea level pressure* SLP, which is adjusted to sea level by a different method, in the remarks section, not in the internationally transmitted part of the code, in hectopascals or millibars. However, in Canada's public weather reports, sea level pressure is instead reported in kilopascals.

In the US weather code remarks, three digits are all that are transmitted; decimal points and the one or two most significant digits are omitted: 1013.2 mbar (101.32 kPa) is transmitted as 132; 1000.0 mbar (100.00 kPa) is transmitted as 000; 998.7 mbar is transmitted as 987; etc. The highest *sea-level pressure* on Earth occurs in Siberia, where the Siberian High often attains a *sea-level pressure* above 1050 mbar (105 kPa; 31 inHg), with record highs close to 1085 mbar (108.5 kPa; 32.0 inHg). The lowest measurable *sea-level pressure* is found at the centers of tropical cyclones and tornadoes, with a record low of 870 mbar (87 kPa; 26 inHg).

Altitude Variation

Pressure varies smoothly from the Earth's surface to the top of the mesosphere. Although the pressure changes with the weather, NASA has averaged the conditions for all parts of the earth year-round. As altitude increases, atmospheric pressure decreases. One can calculate the atmospheric pressure at a given altitude. Temperature and humidity also affect the atmospheric pressure, and it is necessary to know these to compute an accurate figure. The graph at right was developed for a temperature of 15°C and a relative humidity of 0%.

A very local storm above Snæfellsjökull, showing clouds formed on the mountain by orographic lift.

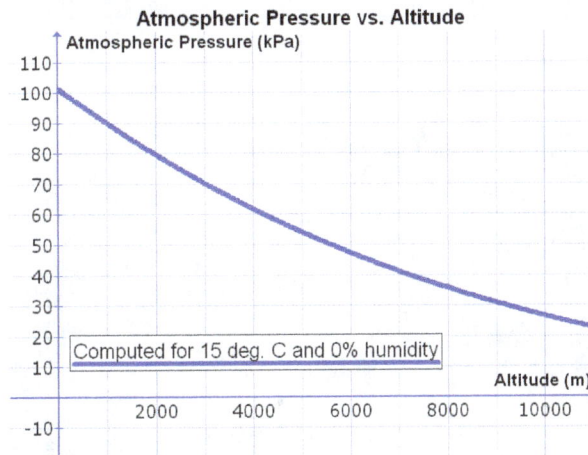

Variation in atmospheric pressure with altitude, computed for 15°C and 0% relative humidity.

This plastic bottle was sealed at approximately 14,000 feet (4,300 m) altitude, and was crushed by the increase in atmospheric pressure —at 9,000 feet (2,700 m) and 1,000 feet (300 m)— as it was brought down towards sea level.

At low altitudes above the sea level, the pressure decreases by about 1.2 kPa for every 100 metres. For higher altitudes within the troposphere, the following equation (the barometric formula) relates atmospheric pressure p to altitude h

$$p = p_0 \cdot \left(1 - \frac{L \cdot h}{T_0}\right)^{\frac{g \cdot M}{R_0 \cdot L}} \approx p_0 \cdot \left(1 - \frac{g \cdot h}{c_p \cdot T_0}\right)^{\frac{c_p \cdot M}{R_0}},$$

$$p \approx p_0 \cdot \exp\left(-\frac{g \cdot M \cdot h}{R_0 \cdot T_0}\right)$$

where the constant parameters are as described below:

Parameter	Description	Value
p_0	sea level standard atmospheric pressure	101325 Pa
L	temperature lapse rate, = g/c_p for dry air	0.0065 K/m
c_p	constant pressure specific heat	~ 1007 J/(kg•K)
T_0	sea level standard temperature	288.15 K
g	Earth-surface gravitational acceleration	9.80665 m/s²
M	molar mass of dry air	0.0289644 kg/mol
R_0	universal gas constant	8.31447 J/(mol•K)

Local Variation

Hurricane Wilma on 19 October 2005; 882 hPa (12.79 psi) in the storm's eye.

Atmospheric pressure varies widely on Earth, and these changes are important in studying weather and climate.

Atmospheric pressure shows a diurnal or semidiurnal (twice-daily) cycle caused by global atmospheric tides. This effect is strongest in tropical zones, with an amplitude of a few millibars, and almost zero in polar areas. These variations have two superimposed cycles, a circadian (24 h) cycle and semi-circadian (12 h) cycle.

Records

The highest adjusted-to-sea level barometric pressure ever recorded on Earth (above 750 meters) was 1085.7 hPa (32.06 inHg) measured in Tosontsengel, Mongolia on 19 December 2001. The highest adjusted-to-sea level barometric pressure ever recorded (below 750 meters) was at Agata in Evenk Autonomous Okrug, Russia (66°53'N, 93°28'E, elevation: 261 m, 856 ft) on 31 December 1968 of 1083.8 hPa (32.005 inHg). The discrimination is due to the problematic assumptions (assuming a standard lapse rate) associated with reduction of sea level from high elevations.

The Dead Sea, the lowest place on Earth at 430 metres (1,410 ft) below sea level, has a correspondingly high typical atmospheric pressure of 1065 hPa.

The lowest non-tornadic atmospheric pressure ever measured was 870 hPa (0.858 atm; 25.69 inHg), set on 12 October 1979, during Typhoon Tip in the western Pacific Ocean. The measurement was based on an instrumental observation made from a reconnaissance aircraft.

Measurement Based on Depth of Water

One atmosphere (101 kPa or 14.7 psi) is the pressure caused by the weight of a column of fresh water of approximately 10.3 m (33.8 ft). Thus, a diver 10.3 m underwater experiences a pressure of about 2 atmospheres (1 atm of air plus 1 atm of water). Conversely, 10.3 m is the maximum height to which water can be raised using suction under standard atmospheric conditions.

Low pressures such as natural gas lines are sometimes specified in inches of water, typically written as *w.c.* (water column) or *w.g.* (inches water gauge). A typical gas-using residential appliance in the US is rated for a maximum of 14 w.c., which is approximately 35 hPa. Similar metric units with a wide variety of names and notation based on millimetres, centimetres or metres are now less commonly used.

Boiling Point of Water

Boiling water

Pure water boils at 100°C (212°F) at earth's standard atmospheric pressure. The boiling point is the temperature at which the vapor pressure is equal to the atmospheric pressure around the water. Because of this, the boiling point of water is lower at lower pressure and higher at higher pressure. Cooking at high elevations, therefore, requires adjustments to recipes. A rough approximation of elevation can be obtained by measuring the temperature at which water boils; in the mid-19th century, this method was used by explorers.

Measurement and Maps

An important application of the knowledge that atmospheric pressure varies directly with altitude was in determining the height of hills and mountains thanks to the availability of reliable pressure measurement devices. While in 1774 Maskelyne was confirming Newton's theory of gravitation at and on Schiehallion in Scotland (using plumb bob deviation to show the effect of "gravity") and accurately measure elevation, William Roy using barometric pressure was able to confirm his height determinations, the agreement being to within one meter (3.28 feet). This was then a useful tool for survey work and map making and long has continued to be useful. It was part of the "application of science" which gave practical people the insight that applied science could easily and relatively cheaply be "useful".

Atmospheric Temperature

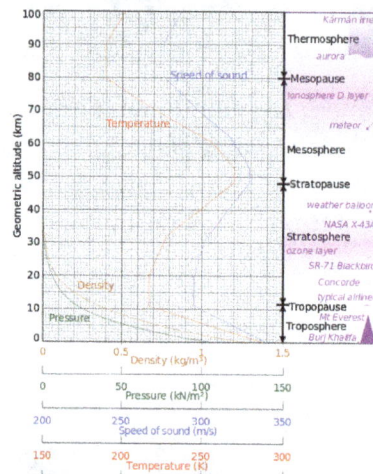

Comparison of the 1962 US Standard Atmosphere graph of geometric altitude against air density, pressure, the speed of sound and temperature with approximate altitudes of various objects.

Atmospheric temperature is a measure of temperature at different levels of the Earth's atmosphere. It is governed by many factors, including incoming solar radiation, humidity and altitude. When discussing surface temperature, the annual atmospheric temperature range at any geographical location depends largely upon the type of biome, as measured by the Köppen climate classification.

Temperature Versus Height

In the Earth's atmosphere, temperature varies greatly at different heights relative to the Earth's surface. The coldest temperatures lie near the mesopause, an area approximately 85 km (53 mi) to 100 km (62 mi) above the surface. In contrast, some of the warmest temperatures can be found in the thermosphere, which receives strong ionizing radiation at the level of the Van Allen radiation belt.

Temperature varies as one moves vertically upwards from the Earth's Surface.

Global Temperature

The concept of a global temperature is commonly used in climatology, and denotes the average temperature of the Earth based on surface, near-surface or tropospheric measurements. These temperature records and measurements are typically acquired using the satellite or ground instrumental temperature measurements, then usually compiled using a database or computer model. Long-term global temperatures in paleoclimate are discerned using proxy data.

Density of Air

The density of air ρ (Greek: rho) (air density) is the mass per unit volume of Earth's atmosphere. Air density, like air pressure, decreases with increasing altitude. It also changes with variation in temperature and humidity. At sea level and at 15°C air has a density of approximately 1.225 kg/m³ (1225.0 g/m³, 0.0023769 slug/(cu ft), 0.0765 lb/(cu ft)) according to ISA (International Standard Atmosphere).

Air density is a property used in many branches of science, engineering, and industry, including aeronautics; gravimetric analysis; the air-conditioning industry; atmospheric research and meteorology; agricultural engineering (modeling and tracking of Soil-Vegetation-Atmosphere-Transfer (SVAT) models); and the engineering community that deals with compressed air.

Density of Air Calculations

Depending on the measuring instruments, use, and necessary rigor of the result, different sets of equations for the calculation of the density of air are used. Air is a mixture of gases and the calculations always simplify, to a greater or lesser extent, the properties of the mixture.

Density of Air Variables

Temperature and Pressure

The density of dry air can be calculated using the ideal gas law, expressed as a function of temperature and pressure:

$$\rho = \frac{p}{R_{specific}T}$$

where:

ρ = air density (kg/m³)

p = absolute pressure (Pa)

T = absolute temperature (K)

$R_{specific}$ = specific gas constant for dry air (J/(kg·K))

The specific gas constant for dry air is 287.058 J/(kg·K) in SI units, and 53.35 (ft·lbf)/(lb·°R) in United States customary and Imperial units. This quantity may vary slightly depending on the molecular composition of air at a particular location.

Therefore:

- At IUPAC standard temperature and pressure (0 °C and 100 kPa), dry air has a density of 1.2754 kg/m³.

- At 20°C and 101.325 kPa, dry air has a density of 1.2041 kg/m³.

- At 70 °F and 14.696 psi, dry air has a density of 0.074887 lb/ft³.

The following table illustrates the air density–temperature relationship at 1 atm or 101.325 kPa:

Effect of temperature on properties of air			
Temperature T (°C)	Speed of sound c (m/s)	Density of air ρ (kg/m³)	Characteristic specific acoustic impedance z_0 (Pa·s/m)
35	351.88	1.1455	403.2
30	349.02	1.1644	406.5
25	346.13	1.1839	409.4
20	343.21	1.2041	413.3
15	340.27	1.2250	416.9
10	337.31	1.2466	420.5
5	334.32	1.2690	424.3
0	331.30	1.2922	428.0
−5	328.25	1.3163	432.1
−10	325.18	1.3413	436.1
−15	322.07	1.3673	440.3
−20	318.94	1.3943	444.6
−25	315.77	1.4224	449.1

Humidity (Water Vapor)

The addition of water vapor to air (making the air humid) reduces the density of the air, which may at first appear counter-intuitive. This occurs because the molar mass of water (18 g/mol) is less than the molar mass of dry air (around 29 g/mol). For any gas, at a given temperature and pressure, the number of molecules present is constant for a particular volume. So when water molecules (water vapor) are added to a given volume of air, the dry air molecules must decrease by the same number, to keep the pressure or temperature from increasing. Hence the mass per unit volume of the gas (its density) decreases.

The density of humid air may be calculated by treating it as a mixture of ideal gases. In this case, the partial pressure of water vapor is known as the vapor pressure. Using this method, error in the density calculation is less than 0.2% in the range of −10°C to 50°C. The density of humid air is found by:

$$\rho_{\text{humid air}} = \frac{p_d}{R_d T} + \frac{p_v}{R_v T} = \frac{p_d M_d + p_v M_v}{RT}$$

where:

$\rho_{\text{humid air}}$ = Density of the humid air (kg/m³)

p_d = Partial pressure of dry air (Pa)

R_d = Specific gas constant for dry air, 287.058 J/(kg·K)

T = Temperature (K)

p_v = Pressure of water vapor (Pa)

R_v = Specific gas constant for water vapor, 461.495 J/(kg·K)

M_d = Molar mass of dry air, 0.028964 kg/mol

M_v = Molar mass of water vapor, 0.018016 kg/mol

R = Universal gas constant, 8.314 J/(K·mol)

The movement of the helicopter rotor leads to a difference in pressure between the upper and lower blade surfaces, allowing the helicopter to fly. A consequence of the pressure change is local variation in air density, strongest in the boundary layer or at transonic speeds.

The vapor pressure of water may be calculated from the saturation vapor pressure and relative humidity. It is found by:

$$p_v = \phi p_{sat}$$

where:

p_v = Vapor pressure of water

ϕ = Relative humidity

p_{sat} = Saturation vapor pressure

The saturation vapor pressure of water at any given temperature is the vapor pressure when relative humidity is 100%. One formula used to find the saturation vapor pressure is:

$$p_{sat} = 6.1078 \times 10^{\frac{7.5T}{T+237.3}}$$

where T = is in degrees C.

note:

- This equation will give the result of pressure in hPa (100 Pa, equivalent to the older unit millibar, 1 mbar = 0.001 bar = 0.1 kPa)

The partial pressure of dry air p_d is found considering partial pressure, resulting in:

$$p_d = p - p_v$$

Where p simply denotes the observed absolute pressure.

Altitude

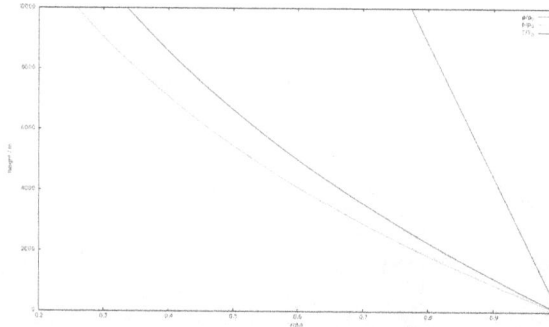

Standard Atmosphere: p_o = 101.325 kPa, T_o = 288.15 K, ρ_o = 1.225 kg/m³

To calculate the density of air as a function of altitude, one requires additional parameters. They are listed below, along with their values according to the International Standard Atmosphere, using for calculation the universal gas constant instead of the air specific constant:

p_0 = sea level standard atmospheric pressure, 101.325 kPa

T_0 = sea level standard temperature, 288.15 K

g = earth-surface gravitational acceleration, 9.80665 m/s²

L = temperature lapse rate, 0.0065 K/m

R = ideal (universal) gas constant, 8.31447 J/(mol·K)

M = molar mass of dry air, 0.0289644 kg/mol

Temperature at altitude h meters above sea level is approximated by the following formula (only valid inside the troposphere):

$$T = T_0 - Lh$$

The pressure at altitude h is given by:

$$p = p_0 \left(1 - \frac{Lh}{T_0} \right)^{\frac{gM}{RL}}$$

Density can then be calculated according to a molar form of the ideal gas law:

$$\rho = \frac{pM}{RT}$$

where:

M = molar mass

R = ideal gas constant

T = absolute temperature

p = absolute pressure

Composition

Composition of dry atmosphere, by volume								
Gas (and others) ppmv		Volume by various		Volume by CIPM-2007		Volume by ASHRAE		
		percentile	ppmv	percentile	ppmv	percentile		
Nitrogen	(N_2)	780,800	(78.080%)	780,848	(78.0848%)	780,818	(78.0818%)	
Oxygen	(O_2)	209,500	(20.950%)	209,390	(20.9390%)	209,435	(20.9435%)	
Argon	(Ar)	9,340	(0.9340%)	9,332	(0.9332%)	9,332	(0.9332%)	
Carbon dioxide	(CO_2)	397.8	(0.03978%)	400	(0.0400%)	385	(0.0385%)	
Neon	(Ne)	18.18	(0.001818%)	18.2	(0.00182%)	18.2	(0.00182%)	
Helium	(He)	5.24	(0.000524%)	5.2	(0.00052%)	5.2	(0.00052%)	
Methane	(CH_4)	1.81	(0.000181%)	1.5	(0.00015%)	1.5	(0.00015%)	
Krypton	(Kr)	1.14	(0.000114%)	1.1	(0.00011%)	1.1	(0.00011%)	▼
Hydrogen	(H_2)	0.55	(0.000055%)	0.5	(0.00005%)	0.5	(0.00005%)	Tap
Nitrous oxide	(N_2O)	0.325	(0.0000325%)	0.3	(0.00003%)	0.3	(0.00003%)	this
Carbon monoxide	(CO)	0.1	(0.00001%)	0.2	(0.00002%)	0.2	(0.00002%)	text
Xenon	(Xe)	0.09	(0.000009%)	0.1	(0.00001%)	0.1	(0.00001%)	
Nitrogen dioxide	(NO_2)	0.02	(0.000002%)	-	-	-	-	to
Iodine	(I_2)	0.01	(0.000001%)	-	-	-	-	expand
Ammonia	(NH_3)	trace	trace	-	-	-	-	or
Sulphur dioxide	(SO_2)	trace	trace	-	-	-	-	collapse
Ozone	(O_3)	0.02 to 0.07	(2 to 7×10^{-6}%)	-	-	-	-	the
Trace to 30 ppm	(----)	-	-	-	-	2.9	(0.00029%)	table
Dry air total	(air)	1,000,065.265	(100.0065265%)	999,997.100	(99.9997100%)	1,000,000.000	(100.0000000%)	▲
Not included in above dry atmosphere:								
Water vapor	(H_2O)	~0.25% by mass over full atmosphere, locally 0.001%–5% by volume.						

- Concentration pertains to the troposphere
- The NASA total value do not add up to exactly 100% due to roundoff and uncertainty. To normalize, N_2 should be reduced by about 51.46 ppmv and O_2 by about 13.805 ppmv.
- ppmv: parts per million by volume (note: volume fraction is equal to mole fraction for ideal gas only
- values disregarded for the calculation of total dry air
- (O_3) concentration up to 0.07 ppmv (7×10^{-6}%) in summer and up to 0.02 ppmv (2×10^{-6}%) in winter
- volumetric composition value adjustment factor (sum of all trace gases, below the (CO_2), and adjusts for 30 ppmv)

Sunlight

Sunlight shining through clouds, giving rise to crepuscular rays

Sunlight is a portion of the electromagnetic radiation given off by the Sun, in particular infrared, visible, and ultraviolet light. On Earth, sunlight is filtered through Earth's atmosphere, and is obvious as daylight when the Sun is above the horizon. When the direct solar radiation is not blocked by clouds, it is experienced as sunshine, a combination of bright light and radiant heat. When it is blocked by clouds or reflects off other objects, it is experienced as diffused light. The World Meteorological Organization uses the term "sunshine duration" to mean the cumulative time during which an area receives direct irradiance from the Sun of at least 120 watts per square meter. Other sources indicate an "Average over the entire earth" of "164 Watts per square meter over a 24 hour day".

The ultraviolet radiation in sunlight has both positive and negative health effects, as it is both a principal source of vitamin D_3 and a mutagen.

Sunlight takes about 8.3 minutes to reach Earth from the surface of the Sun. A photon starting at the center of the Sun and changing direction every time it encounters a charged particle would take between 10,000 and 170,000 years to get to the surface.

Sunlight is a key factor in photosynthesis, the process used by plants and other autotrophic organisms to convert light energy, normally from the Sun, into chemical energy that can be used to fuel the organisms' activities.

Measurement

Researchers may record sunlight using a sunshine recorder, pyranometer, or pyrheliometer. To calculate the amount of sunlight reaching the ground, both Earth's elliptical orbit and the attenuation by Earth's atmosphere have to be taken into account. The extraterrestrial solar illuminance (E_{ext}), corrected for the elliptical orbit by using the day number of the year (dn), is given to a good approximation by

$$E_{\text{ext}} = E_{\text{sc}} \cdot \left(1 + 0.033412 \cdot \cos\left(2\pi \frac{\text{dn} - 3}{365} \right) \right),$$

where dn=1 on January 1st; dn=32 on February 1st; dn=59 on March 1 (except on leap years, where dn=60), etc. In this formula dn−3 is used, because in modern times Earth's perihelion, the closest approach to the Sun and, therefore, the maximum E_{ext} occurs around January 3 each year. The value of 0.033412 is determined knowing that the ratio between the perihelion (0.98328989 AU) squared and the aphelion (1.01671033 AU) squared should be approximately 0.935338.

The solar illuminance constant (E_{sc}), is equal to 128×10^3 lx. The direct normal illuminance (E_{dn}), corrected for the attenuating effects of the atmosphere is given by:

$$E_{\text{dn}} = E_{\text{ext}} e^{-cm},$$

where c is the atmospheric extinction and m is the relative optical airmass. The atmospheric extinction brings the number of lux down to around 100 000.

The total amount of energy received at ground level from the Sun at the zenith depends on the distance to the Sun and thus on the time of year. It is about 3.3% higher than average in January and 3.3% lower in July. If the extraterrestrial solar radiation is 1367 watts per square meter (the value when the Earth–Sun distance is 1 astronomical unit), then the direct sunlight at Earth's surface when the Sun is at the zenith is about 1050 W/m², but the total amount (direct and indirect from the atmosphere) hitting the ground is around 1120 W/m². In terms of energy, sunlight at Earth's surface is around 52 to 55 percent infrared (above 700 nm), 42 to 43 percent visible (400 to 700 nm), and 3 to 5 percent ultraviolet (below 400 nm). At the top of the atmosphere, sunlight is about 30% more intense, having about 8% ultraviolet (UV), with most of the extra UV consisting of biologically damaging short-wave ultraviolet.

Direct sunlight has a luminous efficacy of about 93 lumens per watt of radiant flux. This is higher than the efficacy (of source) of most artificial lighting (including fluorescent), which means using sunlight for illumination heats up a room less than using most forms of artificial lighting.

Multiplying the figure of 1050 watts per square metre by 93 lumens per watt indicates that bright sunlight provides an illuminance of approximately 98 000 lux (lumens per square meter) on a perpendicular surface at sea level. The illumination of a horizontal surface will be considerably less than this if the Sun is not very high in the sky. Averaged over a day, the highest amount of sunlight on a horizontal surface occurs in January at the South Pole.

Dividing the irradiance of 1050 W/m² by the size of the sun's disk in steradians gives an average radiance of 15.4 MW per square metre per steradian. (However, the radiance at the centre of the sun's disk is somewhat higher than the average over the whole disk due to limb darkening.) Multiplying this by π gives an upper limit to the irradiance which can be focused on a surface using mirrors: 48.5 MW/m².

Composition and Power

The spectrum of the Sun's solar radiation is close to that of a black body with a temperature of about 5,800 K. The Sun emits EM radiation across most of the electromagnetic spectrum. Although the Sun produces gamma rays as a result of the nuclear-fusion process, internal absorption and thermalization convert these super-high-energy photons to lower-energy photons before they reach the Sun's surface and are emitted out into space. As a result, the Sun does not emit gamma rays from this process, but it does emit gamma rays from solar flares. The Sun also emits X-rays, ultraviolet, visible light, infrared, and even radio waves; the only direct signature of the nuclear process is the emission of neutrinos.

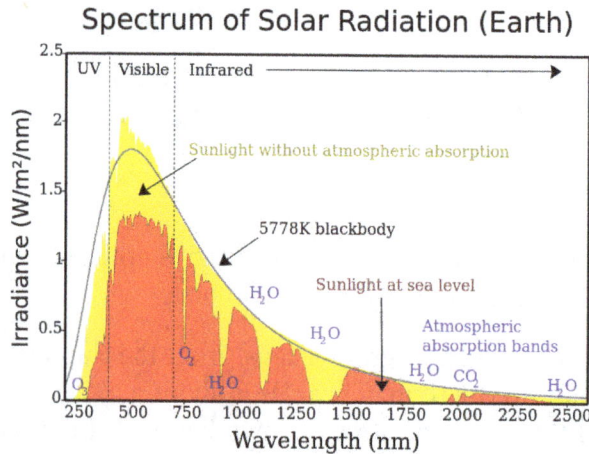

Solar irradiance spectrum above atmosphere and at surface. Extreme UV and X-rays are produced (at left of wavelength range shown) but comprise very small amounts of the Sun's total output power.

Although the solar corona is a source of extreme ultraviolet and X-ray radiation, these rays make up only a very small amount of the power output of the Sun. The spectrum of nearly all solar electromagnetic radiation striking the Earth's atmosphere spans a range of 100 nm to about 1 mm (1,000,000 nm). This band of significant radiation power can be divided into five regions in increasing order of wavelengths:

- Ultraviolet C or (UVC) range, which spans a range of 100 to 280 nm. The term ultraviolet refers to the fact that the radiation is at higher frequency than violet light (and, hence, also invisible to the human eye). Due to absorption by the atmosphere very little reaches Earth's surface. This spectrum of radiation has germicidal properties, as used in germicidal lamps.

- Ultraviolet B or (UVB) range spans 280 to 315 nm. It is also greatly absorbed by the Earth's atmosphere, and along with UVC causes the photochemical reaction leading to the production of the ozone layer. It directly damages DNA and causes sunburn, but is also required for vitamin D synthesis in the skin and fur of mammals.

- Ultraviolet A or (UVA) spans 315 to 400 nm. This band was once held to be less damaging to DNA, and hence is used in cosmetic artificial sun tanning (tanning booths and tanning beds) and PUVA therapy for psoriasis. However, UVA is now known to cause significant damage to DNA via indirect routes (formation of free radicals and reactive oxygen species), and can cause cancer.

- Visible range or light spans 380 to 780 nm. As the name suggests, this range is visible to the naked eye. It is also the strongest output range of the Sun's total irradiance spectrum.

- Infrared range that spans 700 nm to 1,000,000 nm (1 mm). It comprises an important part of the electromagnetic radiation that reaches Earth. Scientists divide the infrared range into three types on the basis of wavelength:

  o Infrared-A: 700 nm to 1,400 nm

  o Infrared-B: 1,400 nm to 3,000 nm

  o Infrared-C: 3,000 nm to 1 mm.

Published Tables

Tables of direct solar radiation on various slopes from 0 to 60 degrees north latitude, in calories per square centimetre, issued in 1972 and published by Pacific Northwest Forest and Range Experiment Station, Forest Service, U.S. Department of Agriculture, Portland, Oregon, USA, appear on the web.

Solar Constant

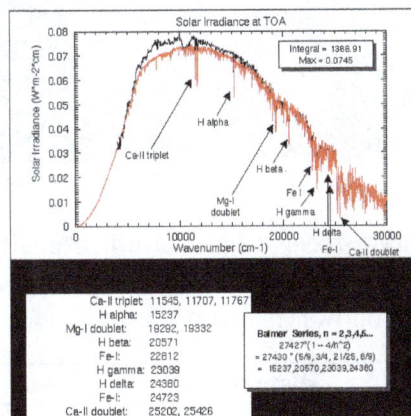

Solar irradiance spectrum at top of atmosphere, on a linear scale and plotted against wavenumber

The solar constant, a measure of flux density, is the amount of incoming solar electromagnetic radiation per unit area that would be incident on a plane perpendicular to the rays, at a distance of one astronomical unit (AU) (roughly the mean distance from the Sun to Earth). The "solar constant" includes all types of solar radiation, not just the visible light. Its average value was thought to be approximately 1366 W/m², varying slightly with solar activity, but recent recalibrations of the relevant satellite observations indicate a value closer to 1361 W/m² is more realistic.

Total Solar Irradiance (TSI) and Spectral Solar Irradiance (SSI) Upon Earth

Total solar irradiance (TSI) – the amount of solar radiation received at the top of Earth's atmosphere – has been measured since 1978 by a series of overlapping NASA and ESA satellite experiments to be 1.361 kilo watts per square meter (kW/m²). TSI observations are continuing today with the ACRIMSAT/ACRIM3, SOHO/VIRGO and SORCE/TIM satellite experiments. Variation of TSI has been discovered on many timescales including the solar magnetic cycle and many shorter periodic cycles. TSI provides the energy that drives Earth's climate, so continuation of the TSI time series database is critical to understanding the role of solar variability in climate change.

Spectral solar irradiance (SSI) – the spectral distribution of the TSI – has been monitored since 2003 by the SORCE Spectral Irradiance Monitor (SIM). It has been found that SSI at UV (ultraviolet) wavelength corresponds in a less clear, and probably more complicated fashion, with Earth's climate responses than earlier assumed, fueling broad avenues of new research in "the connection of the Sun and stratosphere, troposphere, biosphere, ocean, and Earth's climate".

Intensity in the Solar System

Sunlight on Mars is dimmer than on Earth. This photo of a Martian sunset was imaged by Mars Pathfinder.

Different bodies of the Solar System receive light of an intensity inversely proportional to the square of their distance from Sun. A rough table comparing the amount of solar radiation received by each planet in the Solar System follows (from data in):

Planet or dwarf planet	distance (AU)		Solar radiation (W/m²)	
	Perihelion	Aphelion	maximum	minimum
Mercury	0.3075	0.4667	14,446	6,272
Venus	0.7184	0.7282	2,647	2,576
Earth	0.9833	1.017	1,413	1,321
Mars	1.382	1.666	715	492
Jupiter	4.950	5.458	55.8	45.9
Saturn	9.048	10.12	16.7	13.4
Uranus	18.38	20.08	4.04	3.39
Neptune	29.77	30.44	1.54	1.47
Pluto	29.66	48.87	1.55	0.57

The actual brightness of sunlight that would be observed at the surface depends also on the presence and composition of an atmosphere. For example, Venus's thick atmosphere reflects more than 60% of the solar light it receives. The actual illumination of the surface is about 14,000 lux, comparable to that on Earth "in the daytime with overcast clouds".

Sunlight on Mars would be more or less like daylight on Earth during a slightly overcast day, and, as can be seen in the pictures taken by the rovers, there is enough diffuse sky radiation that shadows would not seem particularly dark. Thus, it would give perceptions and "feel" very much like Earth daylight. The spectrum on the surface is slightly redder than that on Earth, due to scattering by reddish dust in the Martian atmosphere.

For comparison purposes, sunlight on Saturn is slightly brighter than Earth sunlight at the average sunset or sunrise. Even on Pluto the sunlight would still be bright enough to almost match the average living room. To see sunlight as dim as full moonlight on Earth, a distance of about 500 AU (~69 light-hours) is needed; there are only a handful of objects in the Solar System known to orbit farther than such a distance, among them 90377 Sedna and (87269) 2000 OO67.

Surface Illumination

The spectrum of surface illumination depends upon solar elevation due to atmospheric effects, with the blue spectral component dominating during twilight before and after sunrise and sunset, respectively, and red dominating during sunrise and sunset. These effects are apparent in natural light photography where the principal source of illumination is sunlight as mediated by the atmosphere.

While the color of the sky is usually determined by Rayleigh scattering, an exception occurs at sunset and twilight. "Preferential absorption of sunlight by ozone over long horizon paths gives the zenith sky its blueness when the sun is near the horizon".

Spectral Composition of Sunlight at Earth's Surface

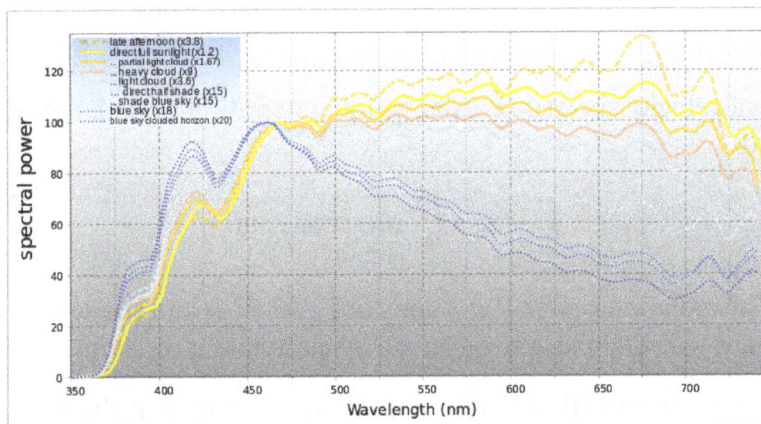

Spectrum of the visible wavelengths at approximately sea level; illumination by direct sunlight compared with direct sunlight scattered by cloud cover and with indirect sunlight by varying degrees of cloud cover. The yellow line shows the spectrum of direct illumination under optimal conditions. The other illumination conditions are scaled to show their relation to direct illumination. The units of spectral power are simply raw sensor values (with a linear response at specific wavelengths).

The Sun's electromagnetic radiation which is received at the Earth's surface is predominantly light that falls within the range of wavelengths to which the visual systems of the animals that inhabit Earth's surface are sensitive. The Sun may therefore be said to illuminate, which is a measure of the light within a specific sensitivity range. Many animals (including humans) have a sensitivity range of approximately 400–700 nm, and given optimal conditions the absorption and scattering by Earth's atmosphere produces illumination that approximates an equal-energy illuminant for most of this range. The useful range for color vision in humans, for example, is approximately 450–650 nm. Aside from effects that arise at sunset and sunrise, the spectral composition changes primarily in respect to how directly sunlight is able to illuminate. When illumination is indirect, Rayleigh scattering in the upper atmosphere will lead blue wavelengths to dominate. Water vapour in the lower atmosphere produces further scattering and ozone, dust and water particles will also absorb selective wavelengths.

Variations in Solar Irradiance

Seasonal and Orbital Variation

On Earth, the solar radiation varies with the angle of the sun above the horizon, with longer sunlight duration at high latitudes during summer, varying to no sunlight at all in winter near the pertinent pole. When the direct radiation is not blocked by clouds, it is experienced as *sunshine*. The warming of the ground (and other objects) depends on the absorption of the electromagnetic radiation in the form of heat.

The amount of radiation intercepted by a planetary body varies inversely with the square of the distance between the star and the planet. Earth's orbit and obliquity change with time (over thousands of years), sometimes forming a nearly perfect circle, and at other times stretching out to an orbital eccentricity of 5% (currently 1.67%). As the orbital eccentricity changes, the average distance from the sun (the semimajor axis does not significantly vary, and so the total insolation over a year remains almost constant due to Kepler's second law,

$$\frac{2A}{r^2} dt = d\theta,$$

where A is the "areal velocity" invariant. That is, the integration over the orbital period (also invariant) is a constant.

$$\int_0^T \frac{2A}{r^2} dt = \int_0^{2\pi} d\theta = \text{constant}.$$

If we assume the solar radiation power P as a constant over time and the solar irradiation given by the inverse-square law, we obtain also the average insolation as a constant.

But the seasonal and latitudinal distribution and intensity of solar radiation received at Earth's surface does vary. The effect of sun angle on climate results in the change in solar energy in summer and winter. For example, at latitudes of 65 degrees, this can vary by more than 25% as a result of Earth's orbital variation. Because changes in winter and summer tend to offset, the change in the annual average insolation at any given location is near zero, but the redistribution of energy between summer and winter does strongly affect the intensity of seasonal cycles. Such changes associated with the redistribution of solar energy are considered a likely cause for the coming and going of recent ice ages.

Solar Intensity Variation

Space-based observations of solar irradiance started in 1978. These measurements show that the solar constant is not constant. It varies on many time scales, including the 11-year sunspot solar cycle. When going further back in time, one has to rely on irradiance reconstructions, using sunspots for the past 400 years or cosmogenic radionuclides for going back 10,000 years. Such reconstructions have been done. These studies show that in addition to the solar irradiance variation with the solar cycle (the (Schwabe) cycle), the solar activitiy varies with longer cycles, such as the proposed 88 year (Gleisberg cycle), 208 year (DeVries cycle) and 1,000 year (Eddy cycle).

Life on Earth

The existence of nearly all life on Earth is fueled by light from the Sun. Most autotrophs, such as plants, use the energy of sunlight, combined with carbon dioxide and water, to produce simple sugars—a process known as photosynthesis. These sugars are then used as building-blocks and in other synthetic pathways that allow the organism to grow.

Heterotrophs, such as animals, use light from the Sun indirectly by consuming the products of autotrophs, either by consuming autotrophs, by consuming their products, or by consuming other heterotrophs. The sugars and other molecular components produced by the autotrophs are then broken down, releasing stored solar energy, and giving the heterotroph the energy required for survival. This process is known as cellular respiration.

In prehistory, humans began to further extend this process by putting plant and animal materials to other uses. They used animal skins for warmth, for example, or wooden weapons to hunt. These skills allowed humans to harvest more of the sunlight than was possible through glycolysis alone, and human population began to grow.

During the Neolithic Revolution, the domestication of plants and animals further increased human access to solar energy. Fields devoted to crops were enriched by inedible plant matter, providing sugars and nutrients for future harvests. Animals that had previously provided humans with only meat and tools once they were killed were now used for labour throughout their lives, fueled by grasses inedible to humans.

The more recent discoveries of coal, petroleum and natural gas are modern extensions of this trend. These fossil fuels are the remnants of ancient plant and animal matter, formed using energy from sunlight and then trapped within Earth for millions of years. Because the stored energy in these fossil fuels has accumulated over many millions of years, they have allowed modern humans to massively increase the production and consumption of primary energy. As the amount of fossil fuel is large but finite, this cannot continue indefinitely, and various theories exist as to what will follow this stage of human civilization (e.g., alternative fuels, Malthusian catastrophe, new urbanism, peak oil).

Cultural Aspects

Many people find direct sunlight to be too bright for comfort, especially when reading from white paper upon which the sun is directly shining. Indeed, looking directly at the sun can cause long-term vision damage. To compensate for the brightness of sunlight, many people wear sunglasses. Cars, many helmets and caps are equipped with visors to block the sun from direct vision when the sun is at a low angle. Sunshine is often blocked from entering buildings through the use of walls, window blinds, awnings, shutters, curtains, or nearby shade trees. The effect of sunlight is relevant to painting, evidenced for instance in works of Claude Monet on outdoor scenes and landscapes.

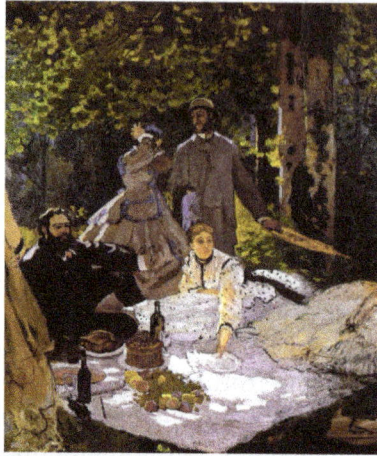

Claude Monet: *Le déjeuner sur l'herbe*

Téli verőfény ("Winter Sunshine") by László Mednyánszky

In colder countries, many people prefer sunnier days and often avoid the shade. In hotter countries, the converse is true; during the midday hours, many people prefer to stay inside to remain cool. If they do go outside, they seek shade that may be provided by trees, parasols, and so on.

In Hinduism, the sun is considered to be a god, as it is the source of life and energy on earth.

Sunbathing

Sunbathing is a popular leisure activity in which a person sits or lies in direct sunshine. People often sunbathe in comfortable places where there is ample sunlight. Some common places for sunbathing include beaches, open air swimming pools, parks, gardens, and sidewalk cafes. Sunbathers typically wear limited amounts of clothing or some simply go nude. For some, an alternative to sunbathing is the use of a sunbed that generates ultraviolet light and can be used indoors regardless of weather conditions. Tanning beds have been banned in a number of states in the world.

For many people with light skin, one purpose for sunbathing is to darken one's skin color (get a sun tan), as this is considered in some cultures to be attractive, associated with outdoor activity, vacations/holidays, and health. Some people prefer naked sunbathing so that an "all-over" or "even" tan can be obtained, sometimes as part of a specific lifestyle.

For people suffering from psoriasis, sunbathing is an effective way of healing the symptoms.

Skin tanning is achieved by an increase in the dark pigment inside skin cells called melanocytes, and is an automatic response mechanism of the body to sufficient exposure to ultraviolet radiation from the sun or from artificial sunlamps. Thus, the tan gradually disappears with time, when one is no longer exposed to these sources.

Effects on Human Health

The ultraviolet radiation in sunlight has both positive and negative health effects, as it is both a principal source of vitamin D_3 and a mutagen. A dietary supplement can supply vitamin D without this mutagenic effect, but bypasses natural mechanisms that would prevent overdoses of vitamin D generated internally from sunlight. Vitamin D has a wide range of positive health effects, which include strengthening bones and possibly inhibiting the growth of some cancers. Sun exposure has also been associated with the timing of melatonin synthesis, maintenance of normal circadian rhythms, and reduced risk of seasonal affective disorder.

Long-term sunlight exposure is known to be associated with the development of skin cancer, skin aging, immune suppression, and eye diseases such as cataracts and macular degeneration. Short-term overexposure is the cause of sunburn, snow blindness, and solar retinopathy.

UV rays, and therefore sunlight and sunlamps, are the only listed carcinogens that are known to have health benefits, and a number of public health organizations state that there needs to be a balance between the risks of having too much sunlight or too little. There is a general consensus that sunburn should always be avoided.

Epidemiological data shows that people who have more exposure to the sun have less high blood pressure and cardiovascular-related mortality. While sunlight (and its UV rays) are a risk factor for skin cancer, "sun avoidance may carry more of a cost than benefit for over-all good health." A study found that there is no evidence that UV reduces lifespan in contrast to other risk factors like smoking, alcohol and high blood pressure.

References

- Chris Landsea (2010-04-21). "Subject: E1), Which is the most intense tropical cyclone on record?". Atlantic Oceanographic and Meteorological Laboratory. Archived from the original on 6 December 2010. Retrieved 2010-11-23

- Berberan-Santos, M. N.; Bodunov, E. N.; Pogliani, L. (1997). "On the barometric formula". American Journal of Physics. 65 (5): 404–412. Bibcode:1997AmJPh..65..404B. doi:10.1119/1.18555

- International Civil Aviation Organization. Manual of the ICAO Standard Atmosphere, Doc 7488-CD, Third Edition, 1993. ISBN 92-9194-004-6

- A., Picard, R.S., Davis, M., Gläser and K., Fujii (2008), Revised formula for the density of moist air (CIPM-2007), Metrologia 45 (2008) 149–155 doi:10.1088/0026-1394/45/2/004, pg 151 Table 1

- "Chapter 8 – Measurement of sunshine duration" (PDF). CIMO Guide. World Meteorological Organization. Retrieved 2008-12-01

- Pollacco, J. A., and B. P. Mohanty (2012), Uncertainties of Water Fluxes in Soil-Vegetation-Atmosphere Transfer Models: Inverting Surface Soil Moisture and Evapotranspiration Retrieved from Remote Sensing, Vadose Zone Journal, 11(3), doi:10.2136/vzj2011.0167

- ICAO, Manual of the ICAO Standard Atmosphere (extended to 80 kilometres (262 500 feet)), Doc 7488-CD, Third Edition, 1993, ISBN 92-9194-004-6

- Wacker M, Holick, MF (2013). "Sunlight and Vitamin D: A global perspective for health.". Dermato-Endocrinology. 5 (1): 51–108. PMC 3897598. PMID 24494042. doi:10.4161/derm.24494

- "NASA: The 8-minute travel time to Earth by sunlight hides a thousand-year journey that actually began in the core". NASA, sunearthday.nasa.gov. Retrieved 2012-02-12

- Shin, Y., B. P. Mohanty, and A.V.M. Ines (2013), Estimating Effective Soil Hydraulic Properties Using Spatially Distributed Soil Moisture and Evapotranspiration, Vadose Zone Journal, 12(3), doi:10.2136/vzj2012.0094

- Marshall,John and Plumb,R. Alan (2008), Atmosphere, ocean, and climate dynamics: an introductory text ISBN 978-0-12-558691-7

- Naylor, Mark; Kevin C. Farmer (1995). "Sun damage and prevention". Electronic Textbook of Dermatology. The Internet Dermatology Society. Retrieved 2008-06-02

- G. Kopp, Greg; J. Lean (2011). "A new, lower value of total solar irradiance: Evidence and climate significance". Geophys. Res. Lett. 38: L01706. Bibcode:2011GeoRL..3801706K. doi:10.1029/2010GL045777

- Wang; et al. (2005). "Modeling the Sun's Magnetic Field and Irradiance since 1713". The Astrophysical Journal. 625 (1): 522–538. Bibcode:2005ApJ...625..522W. doi:10.1086/429689

- ICAO, Manual of the ICAO Standard Atmosphere (extended to 80 kilometres (262 500 feet)), Doc 7488-CD, Third Edition, (1993), ISBN 92-9194-004-6. pg E-x Table B

- "Graph of variation of seasonal and latitudinal distribution of solar radiation". Museum.state.il.us. 2007-08-30. Retrieved 2012-02-12

- Vieira; et al. (2011). "Evolution of the solar irradiance during the Holocene". Astronomy & Astrophysics. 531: A6. Bibcode:2011A&A...531A...6V. arXiv:1103.4958. doi:10.1051/0004-6361/201015843

- Wallace, John M. and Peter V. Hobbs. Atmospheric Science; An Introductory Survey.Elsevier. Second Edition, 2006. ISBN 978-0-12-732951-2

- "13th Report on Carcinogens: Ultraviolet-Radiation-Related Exposures" (PDF). National Toxicology Program. October 2014. Retrieved 2014-12-22

- Osborne JE; Hutchinson PE (August 2002). "Vitamin D and systemic cancer: is this relevant to malignant melanoma?". Br. J. Dermatol. 147 (2): 197–213. PMID 12174089. doi:10.1046/j.1365-2133.2002.04960.x

- Qiang, Fu (2003). "Radiation (Solar)" (PDF). In Holton, James R. Encyclopedia of atmospheric sciences. 5. Amsterdam: Academic Press. pp. 1859–1863. ISBN 978-0-12-227095-6. OCLC 249246073

- Calculated from data in "Reference Solar Spectral Irradiance: Air Mass 1.5". National Renewable Energy Laboratory. Archived from the original on September 28, 2013. Retrieved 2009-11-12

- Egan K; Sosman J; Blot W (February 2, 2005). "Sunlight and Reduced Risk of Cancer: Is The Real Story Vitamin D?". J Natl Cancer Inst. 97 (3): 161–163. doi:10.1093/jnci/dji047

An Overview of Atmospheric Circulation

The atmospheric circulation of our planet differs every year. Polar vortex, Walker circulation and El Niño–Southern Oscillation are some of the topics important to the subject of atmospheric circulation. The major components of atmospheric circulation are discussed in this chapter.

Atmospheric Circulation

Atmospheric circulation is the large-scale movement of air, and together with ocean circulation is the means by which thermal energy is redistributed on the surface of the Earth.

The Earth's atmospheric circulation varies from year to year, but the large scale structure of its circulation remains fairly constant. The smaller scale weather systems – mid-latitude depressions, or tropical convective cells – occur "randomly", and long range weather predictions of those cannot be made beyond ten days in practice, or a month in theory.

The Earth's weather is a consequence of its illumination by the Sun, and the laws of thermodynamics. The atmospheric circulation can be viewed as a heat engine driven by the Sun's energy, and whose energy sink, ultimately, is the blackness of space. The work produced by that engine causes the motion of the masses of air and in that process it redistributes the energy absorbed by the Earth's surface near the tropics to space and incidentally to the latitudes nearer the poles.

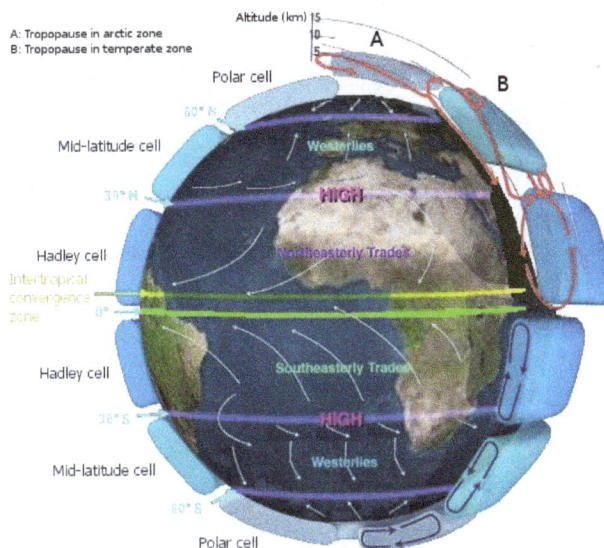

Idealised depiction (at equinox) of large scale atmospheric circulation on Earth.

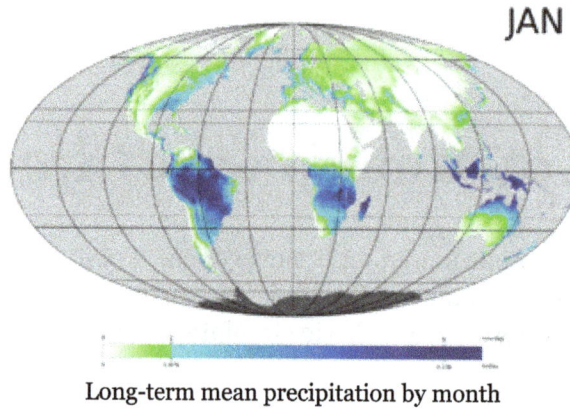

JAN

Long-term mean precipitation by month

The large scale atmospheric circulation "cells" shift polewards in warmer periods (for example, interglacials compared to glacials), but remain largely constant as they are, fundamentally, a property of the Earth's size, rotation rate, heating and atmospheric depth, all of which change little. Over very long time periods (hundreds of millions of years), a tectonic uplift can significantly alter their major elements, such as the jet stream, and plate tectonics may shift ocean currents. During the extremely hot climates of the Mesozoic, a third desert belt may have existed at the Equator.

Latitudinal Circulation Features

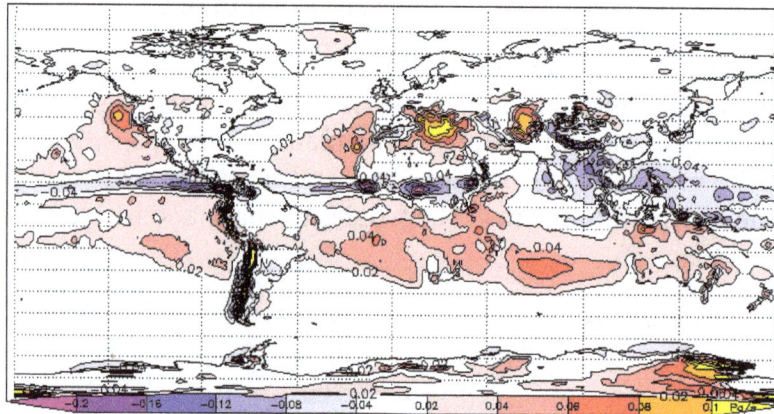

Vertical velocity at 500 hPa, July average. Ascent (negative values) is concentrated close to the solar equator; descent (positive values) is more diffuse but also occurs mainly in the Hadley cell.

The wind belts girdling the planet are organised into three cells in each hemisphere: the Hadley cell, the Ferrel cell, and the Polar cell. Those cells exist in both the northern and southern hemispheres. The vast bulk of the atmospheric motion occurs in the Hadley cell. The high pressure systems acting on the Earth's surface are balanced by the low pressure systems elsewhere. As a result, there is a balance of forces acting on the Earth's surface.

Hadley Cell

The atmospheric circulation pattern that George Hadley described was an attempt to explain the trade winds. The Hadley cell is a closed circulation loop which begins at the equator. There, moist air is warmed by the Earth's surface, decreases in density and rises. A similar air mass rising on the other side of the equator forces those rising air masses to move poleward. The rising air creates a low pressure zone near the equator. As the air moves poleward, it cools, becomes more dense, and descends at about the 30th parallel, creating a high-pressure area. The descended air then travels toward the equator along the surface, replacing the air that rose from the equatorial zone, closing the loop of the Hadley cell. The poleward movement of the air in the upper part of the troposphere deviates toward the east, caused by the coriolis acceleration (a manifestation of conservation of momentum). At the ground level however, the movement of the air toward the equator in the lower troposphere deviates toward the west, producing a wind from the east. The winds that flow to the west (from the east, easterly wind) at the ground level in the Hadley cell are called the Trade Winds.

The ITCZ's band of clouds over the Eastern Pacific and the Americas as seen from space

Though the Hadley cell is described as located at the equator, in the northern hemisphere it shifts to higher latitudes in June and July and toward lower latitudes in December and January, which is the result of the Sun's heating of the surface. The zone where the greatest heating takes place is called the "thermal equator". As the southern hemisphere summer is December to March, the movement of the thermal equator to higher southern latitudes takes place then.

The Hadley system provides an example of a thermally direct circulation. The thermodynamic efficiency and power of the Hadley system, considered as a heat engine, is estimated at 200 terawatts.

Polar Cell

The Polar cell, likewise, is a simple system. Though cool and dry relative to equatorial air, the air masses at the 60th parallel are still sufficiently warm and moist to undergo convection and drive a thermal loop. At the 60th parallel, the air rises to the tropopause (about 8 km at this latitude) and moves poleward. As it does so, the upper level air mass deviates toward the east. When the air reaches the polar areas, it has cooled and is considerably denser than the underlying air. It descends, creating a cold, dry high-pressure area. At the polar surface level, the mass of air is driven toward the 60th parallel, replacing the air that rose there, and the polar circulation cell is complete. As the air at the surface moves toward the equator, it deviates toward the west. Again, the deviations of the air masses are the result of the Coriolis effect. The air flows at the surface are called the polar easterlies.

The outflow of air mass from the cell creates harmonic waves in the atmosphere known as Rossby waves. These ultra-long waves determine the path of the polar jet stream, which travels within the transitional zone between the tropopause and the Ferrel cell. By acting as a heat sink, the polar cell moves the abundant heat from the equator toward the polar regions.

The Hadley cell and the polar cell are similar in that they are thermally direct; in other words, they exist as a direct consequence of surface temperatures. Their thermal characteristics drive the weather in their domain. The sheer volume of energy that the Hadley cell transports, and the depth of the heat sink that is the polar cell, ensures that the effects of transient weather phenomena are not only not felt by the system as a whole, but — except under unusual circumstances — do not form. The endless chain of passing highs and lows which is part of everyday life for mid-latitude dwellers, at latitudes between 30 and 60° latitude, is unknown above the 60th and below the 30th parallels. There are some notable exceptions to this rule. In Europe, unstable weather extends to at least the 70th parallel north.

These atmospheric features are stable. Even though they may strengthen or weaken regionally over time, they do not vanish entirely.

The polar cell, orography and Katabatic winds in Antarctica, can create very cold conditions at the surface, for instance the coldest temperature recorded on Earth: −89.2°C at Vostok Station in Antarctica, measured 1983.

Ferrel Cell

Part of the air rising at 60° latitude diverges at high altitude toward the poles and creates the polar cell. The rest moves toward the equator where it collides at 30° latitude with the high-level air of the Hadley cell. There it subsides and strengthens the high pressure ridges beneath. A large part of the energy that drives the Ferrel cell is provided by the polar and Hadley cells circulating on either side and that drag the Ferrel cell with it. The Ferrel cell, theorized by William Ferrel (1817–1891), is therefore a secondary circulation feature, whose existence depends upon the Hadley and polar cells on either side of it. It might be thought of as an eddy created by the Hadley and polar cells. The Ferrel cell is weak, and the air flow and temperatures within it are variable. For this reason, the mid-latitudes are sometimes known as the "zone of mixing." At high altitudes, the Ferrel cell overrides the Hadley and Polar cells. The air of the Ferrel cell that descends at 30° latitude returns poleward at the ground level, and as it does so it deviates toward the east. In the upper atmosphere of the Ferrel cell, the air moving toward the equator deviates toward the west. Both of those deviations, as in the case of the Hadley and polar cells, are driven by conservation of momentum. As a result, just as the easterly Trade Winds are found below the Hadley cell, the Westerlies are found beneath the Ferrel cell. The forces driving the flow in the Ferrel cell are weak, and so the weather in that zone is variable. Thus, strong high-pressure areas which divert the prevailing westerlies, such as a Siberian high, can override the Ferrel cell, making it discontinuous.

While the Hadley and polar cells are truly closed loops, the Ferrel cell is not, and the telling point is in the Westerlies, which are more formally known as "the Prevailing Westerlies." The easterly Trade Winds and the polar easterlies have nothing over which to prevail, as their parent circulation cells are strong enough and face few obstacles either in the form of massive terrain features or high pressure zones. The weaker Westerlies of the Ferrel cell, however, can be disrupted. The local passage of a cold front may change that in a matter of minutes, and frequently does. As a result, at the surface, winds can vary abruptly in direction. But the winds above the surface, where they are less disrupted by terrain, are essentially westerly. A low pressure zone at 60° latitude that moves toward the equator, or a high pressure zone at 30° latitude that moves poleward, will accelerate the Westerlies of the Ferrel cell. A strong high, moving polewards may bring westerly winds for days.

The Ferrel cell is driven by the Hadley and Polar cells. It has neither a strong source of heat nor a strong sink to drive convection. As a result, the weather within the Ferrel cell is highly variable and is influenced by changes to the Hadley and Polar cells. The base of the Ferrel cell is characterized by the movement of air masses, and the location of those air masses is influenced in part by the location of the jet stream, even though it flows near the tropopause. Overall, the movement of surface air is from the 30th latitude to the 60th. However, the upper flow of the Ferrel cell is weak and not well defined.

In contrast to the Hadley and Polar systems, the Ferrel system provides an example of a thermally indirect circulation. The Ferrel system acts as a heat pump with a coefficient of performance of 12.1, consuming kinetic energy at an approximate rate of 275 terrawatts.

Longitudinal Circulation Features

While the Hadley, Ferrel, and polar cells (whose axes are oriented along parallels or latitudes) are the major features of global heat transport, they do not act alone. Temperature differences also drive a set of circulation cells, whose axes of circulation are longitudinally oriented. This atmospheric motion is known as zonal overturning circulation.

Latitudinal circulation is a result of the highest solar radiation per unit area (solar intensity) falling on the tropics. The solar intensity decreases as the latitude increases, reaching essentially zero at the poles. Longitudinal circulation, however, is a result of the heat capacity of water, its absorptivity, and its mixing. Water absorbs more heat than does the land, but its temperature does not rise as greatly as does the land. As a result, temperature variations on land are greater than on water. The Hadley, Ferrel, and polar cells operate at the largest scale of thousands of kilometers (synoptic scale). But, even at mesoscales (a horizontal range of 5 to several hundred kilometres), this effect is noticeable. During the day, air warmed by the relatively hotter land rises, and as it does so it draws a cool breeze from the sea that replaces the risen air. At night, the relatively warmer water and cooler land reverses the process, and a breeze from the land, of air cooled by the land, is carried offshore by night. This described effect is daily (diurnal).

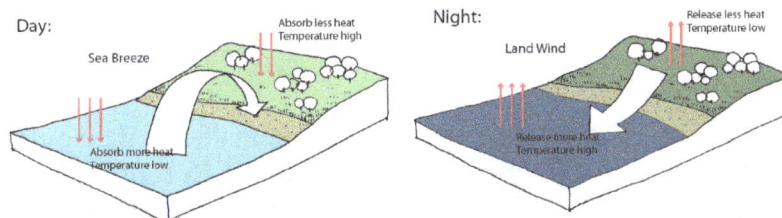

Diurnal wind change in local coastal area, also applies on the continental scale.

At the larger, synoptic, scale of oceans and continents, this effect is seasonal or even decadal. Warm air rises over the equatorial, continental, and western Pacific Ocean regions. When it reaches the tropopause, it cools and subsides in a region of relatively cooler water mass.

The Pacific Ocean cell plays a particularly important role in Earth's weather. This entirely ocean-based cell comes about as the result of a marked difference in the surface temperatures of the western and eastern Pacific. Under ordinary circumstances, the western Pacific waters are warm, and the eastern waters are cool. The process begins when strong convective activity over equatorial East Asia and subsiding cool air off South America's west coast creates a wind pattern which pushes Pacific water westward and piles it up in the western Pacific. (Water levels in the western Pacific are about 60 cm higher than in the eastern Pacific.)

Walker Circulation

The Pacific cell is of such importance that it has been named the Walker circulation after Sir Gilbert Walker, an early-20th-century director of British observatories in India, who sought a means of predicting when the monsoon winds of India would fail. While he was never successful in doing so, his work led him to the discovery of a link between the periodic pressure variations in the Indian Ocean, and those between the eastern and western Pacific, which he termed the "Southern Oscillation".

The movement of air in the Walker circulation affects the loops on either side. Under normal circumstances, the weather behaves as expected. But every few years, the winters become unusually warm or unusually cold, or the frequency of hurricanes increases or decreases, and the pattern sets in for an indeterminate period.

The Walker Cell plays a key role in this and in the El Niño phenomenon. If convective activity slows in the Western Pacific for some reason (this reason is not currently known), the climates of areas adjacent to the Western Pacific are affected. First, the upper-level westerly winds fail. This cuts off the source of returning, cool air that would normally subside at about 30° north latitude, and therefore the air returning as surface easterlies ceases. The consequence of this is twofold. Warm water ceases to surge into the eastern Pacific from the west (it was "piled" by past easterly winds) since there is no longer a surface wind to push it into the area of the west pacific. This and the corresponding effects of the Southern Oscillation result in long-term unseasonable temperatures and precipitation patterns in North and South America, Australia, and Southeast Africa, and the disruption of ocean currents.

Meanwhile, in the Atlantic, fast-blowing upper level Westerlies of the Hadley cell form, which would ordinarily be blocked by the Walker circulation and unable to reach such intensities. These winds disrupt the tops of nascent hurricanes and greatly diminish the number which are able to reach full strength.

El Niño – Southern Oscillation

El Niño and *La Niña* are opposite surface temperature anomalies of the Southern Pacific, which heavily influence the weather on a large scale. In the case of El Niño, warm surface water approaches the coasts of South America which results in blocking the upwelling of nutrient-rich deep water. This has serious impacts on the fish populations.

In the La Niña case, the convective cell over the western Pacific strengthens inordinately, resulting in colder than normal winters in North America and a more robust cyclone season in South-East Asia and Eastern Australia. There is also an increased upwelling of deep cold ocean waters and more intense uprising of surface air near South America, resulting in increasing numbers of drought occurrences, although fishermen reap benefits from the more nutrient-filled eastern Pacific waters.

Polar Vortex

A polar vortex is an upper level low-pressure area lying near the Earth's pole. There are two polar vortices in the Earth's atmosphere, which overlie the North, and South Poles. Each polar vortex is a persistent, large-scale, low pressure zone that rotates counter-clockwise at the North Pole (called a cyclone), and clockwise at the South Pole. The bases of the two polar vortices are located in the middle and upper troposphere and extend into the stratosphere. Beneath that lies a large mass of cold, dense arctic air. The vortices weaken and strengthen from year to year. When the vortex of the arctic is strong it is well defined, there is a single vortex and the arctic air is well contained; when weaker, which it generally is, it will break into two or more vortices; when very weak, the flow of arctic air becomes more disorganized and masses of cold arctic air can push equatorward, bringing with it a rapid and sharp temperature drop. The interface between the cold dry air mass of the pole and the warm moist air mass further south defines the location of the polar front. The polar front is centered, roughly at 60° latitude. A polar vortex strengthens in the winter and weakens in the summer due to its dependence on the temperature difference between the equator and the poles. The vortices span less than 1,000 kilometers (620 miles) in diameter within which they rotate counter-clockwise in the Northern Hemisphere, and in a clockwise fashion in the Southern Hemisphere. As with other cyclones, their rotation is driven by the Coriolis effect.

The Arctic polar vortex

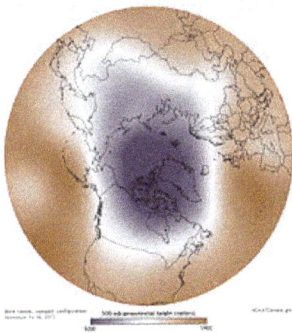

The typical polar vortex configuration in November, 2013.

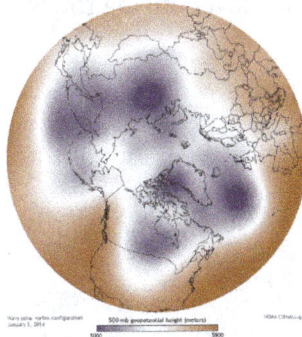

A weak polar vortex on January 5, 2014.

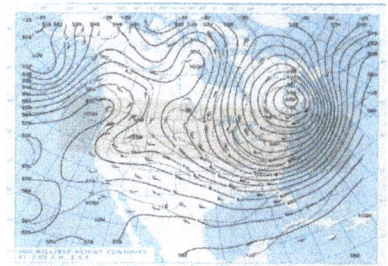

Low pressure area over Quebec and Maine, part of the northern polar vortex weakening, on the record-setting cold morning of January 21, 1985.

When the polar vortex is strong, there is a single vortex with a jet stream that is "well constrained" near the polar front. When the northern vortex weakens, it separates into two or more vortices, the strongest of which are near Baffin Island, Canada and the other over northeast Siberia. The Antarctic vortex of the Southern Hemisphere is a single low pressure zone that is found near the edge of the Ross ice shelf near 160 west longitude. When the polar vortex is strong, the mid-latitude Westerlies (winds at the surface level between 30° and 60° latitude from the west) increase in strength and are persistent. When the polar vortex is weak, high pressure zones of the mid latitudes may push poleward, moving the polar vortex, jet stream, and polar front equatorward. The jet stream is seen to "buckle" and deviate south. This rapidly brings cold dry air into contact with the warm, moist air of the mid latitudes, resulting in a rapid and dramatic change of weather known as a "cold snap".

Ozone depletion occurs within the polar vortices – particularly over the Southern Hemisphere – reaching a maximum depletion in the spring.

History

The polar vortex was first described as early as 1853. The phenomenon's sudden stratospheric warming (SSW) develops during the winter in the Northern Hemisphere and was discovered in 1952 with radiosonde observations at altitudes higher than 20 km.

The phenomenon was mentioned frequently in the news and weather media in the cold North American winter of 2013-2014, popularizing the term as an explanation of very cold temperatures.

Identification

Polar cyclones are low pressure zones embedded within the polar air masses, and exist year-round. The stratospheric polar vortex develops at latitudes above the subtropical jet stream. Horizontally, most polar vortices have a radius of less than 1,000 kilometres (620 mi). Since polar vortices exist from the stratosphere downward into the mid-troposphere, a variety of heights/pressure levels are used to mark its position. The 50 mb pressure surface is most often used to identify its stratospheric location. At the level of the tropopause, the extent of closed contours of potential temperature can be used to determine its strength. Others have used levels down to the 500 hPa pressure level (about 5,460 metres (17,910 ft) above sea level during the winter) to identify the polar vortex.

Duration and Power

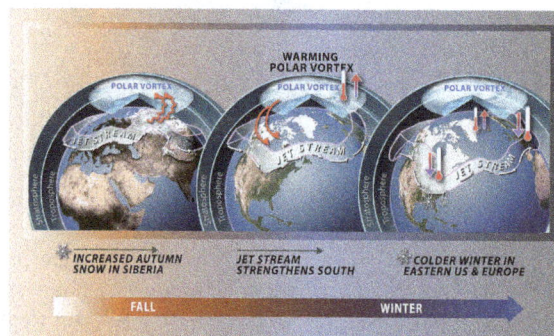

Polar vortex and weather impacts due to stratospheric warming

Polar vortices are weakest during summer and strongest during winter. Extratropical cyclones that migrate into higher latitudes when the polar vortex is weak can disrupt the single vortex creating smaller vortices (cold-core lows) within the polar air mass. Those individual vortices can persist for more than a month.

Volcanic eruptions in the tropics can lead to a stronger polar vortex during winter for as long as two years afterwards. The strength and position of the polar vortex shapes the flow pattern in a broad area about it. An index which is used in the northern hemisphere to gauge its magnitude is the Arctic oscillation.

When the Arctic vortex is at its strongest, there is a single vortex, but normally, the Arctic vortex is elongated in shape, with two cyclone centers, one over Baffin Island in Canada and the other over northeast Siberia. When the Arctic pattern is at its weakest, subtropic air masses can intrude poleward causing the Arctic air masses to move equatorward, as during the Winter 1985 Arctic outbreak. The Antarctic polar vortex is more pronounced and persistent than the Arctic one. In the Arctic the distribution of land masses at high latitudes in the Northern Hemisphere gives rise to Rossby waves which contribute to the breakdown of the polar vortex, whereas in the Southern Hemisphere the vortex is less disturbed. The breakdown of the polar vortex is an extreme event known as a sudden stratospheric warming, here the vortex completely breaks down and an associated warming of 30–50°C (54–90°F) over a few days can occur.

The waxing and waning of the polar vortex is driven by the movement of mass and the transfer of heat in the polar region. In the autumn, the circumpolar winds increase in speed and the polar vortex rises into the stratosphere. The result is the polar air forms a coherent rotating air mass: the polar vortex. As winter approaches, the vortex core cools, the winds decrease, and the vortex energy declines. Once late winter and early spring approach the vortex is at its weakest. As a result, during late winter, large fragments of the vortex air can be diverted into lower latitudes by stronger weather systems intruding from those latitudes. In the lowest level of the stratosphere, strong potential vorticity gradients remain, and the majority of that air remains confined within the polar air mass into December in the Southern Hemisphere and April in the Northern Hemisphere, well after the breakup of the vortex in the mid-stratosphere.

The breakup of the northern polar vortex occurs between mid March to mid May. This event signifies the transition from winter to spring, and has impacts on the hydrological cycle, growing seasons of vegetation, and overall ecosystem productivity. The timing of the transition also influences changes in sea ice, ozone, air temperature, and cloudiness. Early and late polar breakup episodes have occurred, due to variations in the stratospheric flow structure and upward spreading of planetary waves from the troposphere. As a result of increased waves into the vortex, the vortex experiences more rapid warming than normal, resulting in an earlier breakup and spring. When the breakup comes early, it is characterized by with persistent of remnants of the vortex. When the breakup is late, the remnants dissipate rapidly. When the breakup is early, there is one warming period from late February to middle March. When the breakup is late, there are two warming periods, one January, and one in March. Zonal mean temperature, wind, and geopotential height exert varying deviations from their normal values before and after early breakups, while the deviations remain constant before and after late breakups. Scientists are connecting a delay in the Arctic vortex breakup with a reduction of planetary wave activities, few stratospheric sudden warming events, and depletion of ozone.

Sudden stratospheric warming events are associated with weaker polar vortices. This warming of stratospheric air can reverse the circulation in the Arctic Polar Vortex from counter-clockwise to clockwise. These changes aloft force changes in the troposphere below. An example of an effect on the troposphere is the change in speed of the Atlantic Ocean circulation pattern. A soft spot just south of Greenland is where the initial step of downwelling occurs, nicknamed the "Achilles

Heel of the North Atlantic". Small amounts of heating or cooling traveling from the polar vortex can trigger or delay downwelling, altering the Gulf Stream Current of the Atlantic, and the speed of other ocean currents. Since all other oceans depend on the Atlantic Ocean's movement of heat energy, climates across the planet can be dramatically affected. The weakening or strengthening of the polar vortex can alter the sea circulation more than a mile beneath the waves. Strengthening storm systems within the troposphere that cool the poles, intensify the polar vortex. La Niña–related climate anomalies significantly strengthen the polar vortex. Intensification of the polar vortex produces changes in relative humidity as downward intrusions of dry, stratospheric air enter the vortex core. With a strengthening of the vortex comes a longwave cooling due to a decrease in water vapor concentration near the vortex. The decreased water content is a result of a lower tropopause within the vortex, which places dry stratospheric air above moist tropospheric air. Instability is caused when the vortex tube, the line of concentrated vorticity, is displaced. When this occurs, the vortex rings become more unstable and prone to shifting by planetary waves.The planetary wave activity in both hemispheres varies year-to-year, producing a corresponding response in the strength and temperature of the polar vortex. The number of waves around the perimeter of the vortex are related to the core size; as the vortex core decreases, the number of waves increase.

The degree of the mixing of polar and mid-latitude air depends on the evolution and position of the polar night jet. In general, the mixing is less inside the vortex than outside. Mixing occurs with unstable planetary waves that are characteristic of the middle and upper stratosphere in winter. Prior to vortex breakdown, there is little transport of air out of the Arctic Polar Vortex due to strong barriers above 420 km (261 miles). The polar night jet which exists below this, is weak in the early winter. As a result, it does not deviate any descending polar air, which then mixes with air in the mid-latitudes. In the late winter, air parcels do not descend as much, reducing mixing. After the vortex is broken up, the ex-vortex air is dispersed into the middle latitudes within a month.

Sometimes, a mass of the polar vortex breaks off before the end of the final warming period. If large enough, the piece can move into Canada and the Midwestern, Central, Southern, and Northeastern United States. This diversion of the polar vortex can occur due to the displacement of the polar jet stream; for example, the significant northwestward direction of the polar jet stream in the western part of the United States during the winters of 2013–2014, and 2014-2015. This caused warm, dry conditions in the west, and cold, snowy conditions in the east. Occasionally, the high-pressure air mass, called the Greenland Block, can cause the polar vortex to divert to the south, rather than follow its normal path over the North Atlantic.

Climate Change

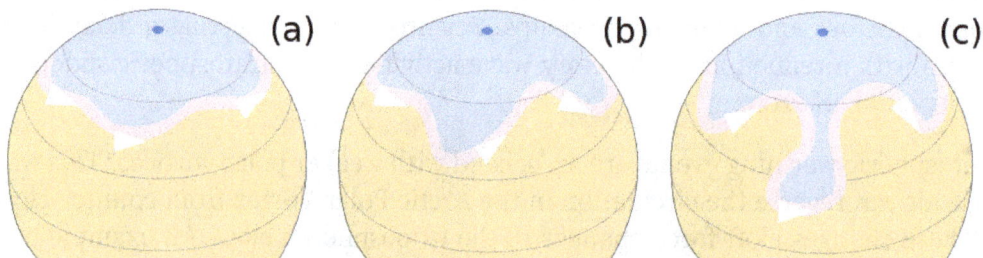

Meanders of the northern hemisphere's jet stream developing (a, b) and finally detaching a "drop" of cold air (c); orange: warmer masses of air; pink: jet stream.

A study in 2001 found that stratospheric circulation can have anomalous effects on weather regimes. In the same year researchers found a statistical correlation between weak polar vortex and outbreaks of severe cold in the Northern Hemisphere. In more recent years scientists identified interactions with Arctic sea ice decline, reduced snow cover, evapotranspiration patterns, NAO anomalies or weather anomalies which are linked to the polar vortex and jet stream configuration. However, because the specific observations are considered short-term observations (starting c. 13 years ago) there is considerable uncertainty in the conclusions. Climatology observations require several decades to definitively distinguish natural variability from climate trends.

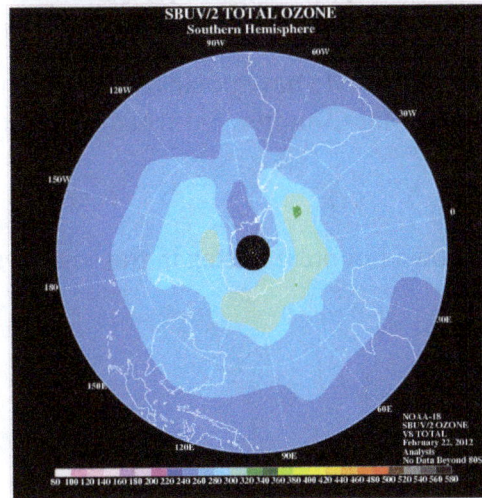

Southern Hemisphere Ozone Concentration, February 22, 2012

The general assumption is that reduced snow cover and sea ice reflect less sunlight and therefore evaporation and transpiration increases, which in turn alters the pressure and temperature gradient of the polar vortex, causing it to weaken or collapse. This becomes apparent when the jet stream amplitude increases (meanders) over the northern hemisphere, causing Rossby waves to propagate farther to the south or north, which in turn transports warmer air to the north pole and polar air into lower latitudes. The jet stream amplitude increases with a weaker polar vortex, hence increases the chance for weather systems to become blocked. A recent blocking event emerged when a high-pressure over Greenland steered Hurricane Sandy into the northern Mid-Atlantic states.

Ozone Depletion

The chemistry of the Antarctic polar vortex has created severe ozone depletion. The nitric acid in polar stratospheric clouds reacts with chlorofluorocarbons to form chlorine, which catalyzes the photochemical destruction of ozone. Chlorine concentrations build up during the polar winter, and the consequent ozone destruction is greatest when the sunlight returns in spring. These clouds can only form at temperatures below about −80°C (−112°F). Since there is greater air exchange between the Arctic and the mid-latitudes, ozone depletion at the north pole is much less severe than at the south. Accordingly, the seasonal reduction of ozone levels over the Arctic is usually characterized as an "ozone dent", whereas the more severe ozone depletion over the Antarctic is considered an "ozone hole". This said, chemical ozone destruction in the 2011 Arctic polar vortex attained, for the first time, a level clearly identifiable as an Arctic "ozone hole".

Outside Earth

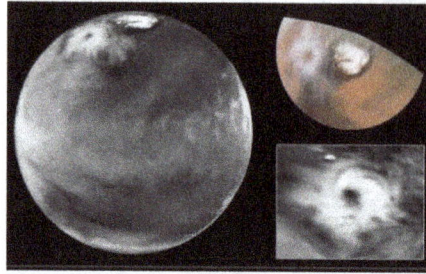

Hubble view of the colossal polar cloud on Mars

Other astronomical bodies are also known to have polar vortices, including Venus (double vortex—that is, two polar vortices at a pole), Mars, Jupiter, Saturn, and Saturn's moon Titan.

Hot Polar Vortex

Saturn's south pole is the only known hot polar vortex in the solar system.

Walker Circulation

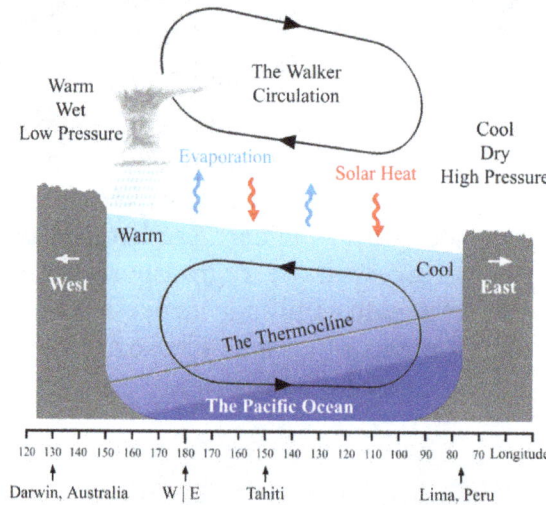

A schematic diagram of the quasi-equilibrium and La Niña phase of the southern oscillation.
The Walker circulation is seen at the surface as easterly trade winds which move water and air warmed by the sun towards the west. The western side of the equatorial Pacific is characterized by warm, wet low pressure weather as the collected moisture is dumped in the form of typhoons and thunderstorms. The ocean is some 60 cm higher in the western Pacific as the result of this motion. The water and air are returned to the east. Both are now much cooler, and the air is much drier. An El Niño episode is characterised by a breakdown of this water and air cycle, resulting in relatively warm water and moist air in the eastern Pacific.

The Walker circulation, also known as the Walker cell, is a conceptual model of the air flow in the tropics in the lower atmosphere (troposphere). According to this model, parcels of air follow a closed circulation in the zonal and vertical directions. This circulation, which is roughly consistent with observations, is caused by differences in heat distribution between ocean and land. It was discovered by Gilbert Walker. In addition to motions in the zonal and vertical direction the tropical atmosphere also has considerable motion in the meridional direction as part of, for example, the Hadley Circulation.

The term "Walker circulation" was coined in 1969 by the Norwegian-American meteorologist Jacob Bjerknes.

Walker's Methodology

Gilbert Walker was an established applied mathematician at the University of Cambridge when he became director-general of observatories in India in 1904. While there, he studied the characteristics of the Indian Ocean monsoon, the failure of whose rains had brought severe famine to the country in 1899. Analyzing vast amounts of weather data from India and the rest of the world, over the next fifteen years he published the first descriptions of the great seesaw oscillation of atmospheric pressure between the Indian and Pacific Ocean, and its correlation to temperature and rainfall patterns across much of the Earth's tropical regions, including India. He also worked with the Indian Meteorological Department especially in linking the monsoon with Southern Oscillation phenomenon. He was made a Companion of the Order of the Star of India in 1911.

Walker determined that the time scale of a year (used by many studying the atmosphere) was unsuitable because geospatial relationships could be entirely different depending on the season. Thus, Walker broke his temporal analysis into December–February, March–May, June–August, and September–November.

Walker then selected a number of "centers of action", which included areas such as the Indian Peninsula. The centers were in the hearts of regions with either permanent or seasonal high and low pressures. He also added points for regions where rainfall, wind or temperature was an important control.

He examined the relationships of the summer and winter values of pressure and rainfall, first focusing on summer and winter values, and later extending his work to the spring and autumn.

He concludes that variations in temperature are generally governed by variations in pressure and rainfall. It had previously been suggested that sunspots could be the cause of the temperature variations, but Walker argued against this conclusion by showing monthly correlations of sunspots with temperature, winds, cloud cover, and rain that were inconsistent.

Walker made it a point to publish all of his correlation findings, both of relationships found to be important as well as relationships that were found to be unimportant. He did this for the purpose of dissuading researchers from focusing on correlations that did not exist.

Oceanic Effects

The Walker Circulations of the tropical Indian, Pacific, and Atlantic basins result in westerly surface winds in Northern Summer in the first basin and easterly winds in the second and third basins. As a result, the temperature structure of the three oceans display dramatic asymmetries. The equatorial Pacific and Atlantic both have cool surface temperatures in Northern Summer in the east, while cooler surface temperatures prevail only in the western Indian Ocean. These changes in surface temperature reflect changes in the depth of the thermocline.

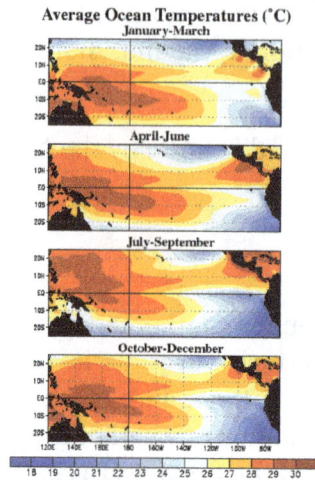

Average equatorial Pacific temperatures

Changes in the Walker Circulation with time occur in conjunction with changes in surface temperature. Some of these changes are forced externally, such as the seasonal shift of the Sun into the Northern Hemisphere in summer. Other changes appear to be the result of coupled ocean-atmosphere feedback in which, for example, easterly winds cause the sea surface temperature to fall in the east, enhancing the zonal heat contrast and hence intensifying easterly winds across the basin. These anomalous easterlies induce more equatorial upwelling and raise the thermocline in the east, amplifying the initial cooling by the southerlies. This coupled ocean-atmosphere feedback was originally proposed by Bjerknes. From an oceanographic point of view, the equatorial cold tongue is caused by easterly winds. Were the earth climate symmetric about the equator, cross-equatorial wind would vanish, and the cold tongue would be much weaker and have a very different zonal structure than is observed today. The Walker cell is indirectly related to upwelling off the coasts of Peru and Ecuador. This brings nutrient-rich cold water to the surface, increasing fishing stocks.

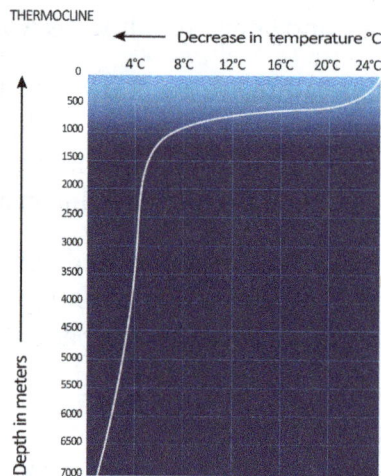

Graph showing a tropical ocean thermocline (depth vs. temperature). Note the rapid change between 100 and 1000 meters. The temperature is nearly constant after 1500 meters depth.

El Niño

The Walker circulation is caused by the pressure gradient force that results from a high pressure system over the eastern Pacific Ocean, and a low pressure system over Indonesia. When the Walker circulation weakens or reverses, an El Niño results, causing the ocean surface to be warmer than average, as upwelling of cold water occurs less or not at all. An especially strong Walker circulation causes a La Niña, resulting in cooler ocean temperatures due to increased upwelling.

A scientific study published in May 2006 in the journal Nature indicates that the Walker circulation has been slowing since the mid-19th Century. The authors argue that global warming is a likely causative factor in the weakening of the wind pattern. However, a new study from The Twentieth Century Reanalysis Project shows that the Walker circulation has not been slowing (or increasing) from 1871–2008.

El Niño–Southern Oscillation

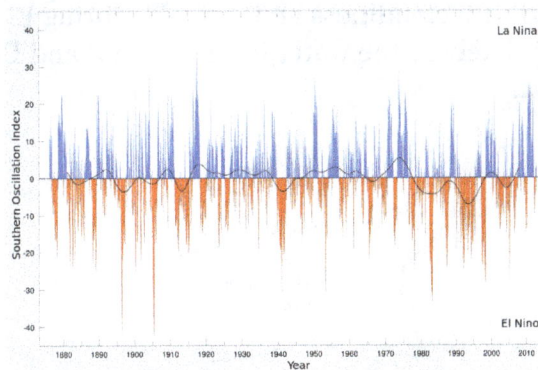

Southern Oscillation Index timeseries 1876–2012

El Niño–Southern Oscillation (*ENSO*) is an irregularly periodical variation in winds and sea surface temperatures over the tropical eastern Pacific Ocean, affecting much of the tropics and subtropics. The warming phase is known as *El Niño* and the cooling phase as *La Niña*. *Southern Oscillation* is the accompanying atmospheric component, coupled with the sea temperature change: *El Niño* is accompanied with high, and *La Niña* with low air surface pressure in the tropical western Pacific. The two periods last several months each (typically occur every few years) and their effects vary in intensity.

The two phases relate to the Walker circulation, discovered by Gilbert Walker during the early twentieth century. The Walker circulation is caused by the pressure gradient force that results from a high pressure system over the eastern Pacific Ocean, and a low pressure system over Indonesia. When the Walker circulation weakens or reverses, an *El Niño* results, causing the ocean surface to be warmer than average, as upwelling of cold water occurs less or not at all. An especially strong Walker circulation causes a *La Niña*, resulting in cooler ocean temperatures due to increased upwelling.

Mechanisms that cause the oscillation remain under study. The extremes of this climate pattern's oscillations cause extreme weather (such as floods and droughts) in many regions of the world. Developing countries dependent upon agriculture and fishing, particularly those bordering the Pacific Ocean, are the most affected.

Concept

The El Niño–Southern Oscillation is a single climate phenomenon that periodically fluctuates between three phases: Neutral, La Niña or El Niño. La Niña and El Niño are opposite phases that require certain changes to take place in both the ocean and the atmosphere, before an event is declared.

Normally the northward flowing Humboldt Current brings relatively cold water from the Southern Ocean northwards along South America's west coast to the tropics, where it is enhanced by up-welling taking place along the coast of Peru. Along the equator trade winds cause the ocean currents in the eastern Pacific to draw water from the deeper ocean towards the surface, helping to keep the surface cool. Under the influence of the equatorial trade winds, this cold water flows westwards along the equator where it is slowly heated by the sun. As a direct result sea surface temperatures in the western Pacific are generally warmer, by about 8–10°C (14–18°F) than those in the Eastern Pacific. This warmer area of ocean is a source for convection and is associated with cloudiness and rainfall. During El Niño years the cold water weakens or disappears completely as the water in the Central and Eastern Pacific becomes as warm as the Western Pacific.

Walker circulation

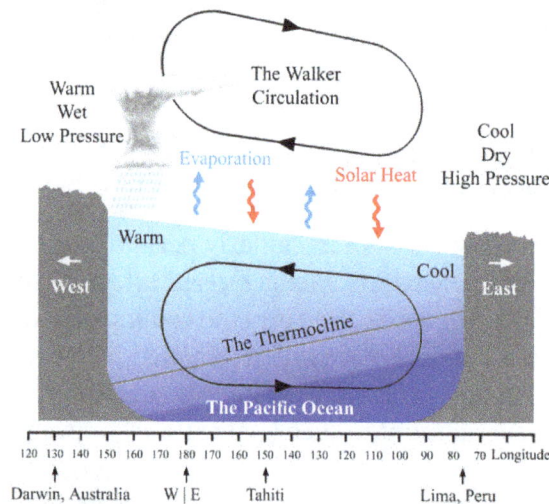

Diagram of the quasi-equilibrium and La Niña phase of the Southern Oscillation. The Walker circulation is seen at the surface as easterly trade winds which move water and air warmed by the sun towards the west. The western side of the equatorial Pacific is characterized by warm, wet low pressure weather as the collected moisture is dumped in the form of typhoons and thunderstorms. The ocean is some 60 centimetres (24 in) higher in the western Pacific as the result of this motion. The water and air are returned to the east. Both are now much cooler, and the air is much drier. An El Niño episode is characterised by a breakdown of this water and air cycle, resulting in relatively warm water and moist air in the eastern Pacific.

The Walker circulation is caused by the pressure gradient force that results from a high pressure system over the eastern Pacific Ocean, and a low pressure system over Indonesia. The Walker circulations of the tropical Indian, Pacific, and Atlantic basins result in westerly surface winds in northern summer in the first basin and easterly winds in the second and third basins. As a result, the temperature structure of the three oceans display dramatic asymmetries. The equatorial Pacific and Atlantic both have cool surface temperatures in northern summer in the east, while cooler surface temperatures prevail only in the western Indian Ocean. These changes in surface temperature reflect changes in the depth of the thermocline.

Changes in the Walker circulation with time occur in conjunction with changes in surface temperature. Some of these changes are forced externally, such as the seasonal shift of the sun into the Northern Hemisphere in summer. Other changes appear to be the result of coupled ocean-atmosphere feedback in which, for example, easterly winds cause the sea surface temperature to fall in the east, enhancing the zonal heat contrast and hence intensifying easterly winds across the basin. These anomalous easterlies induce more equatorial upwelling and raise the thermocline in the east, amplifying the initial cooling by the southerlies. This coupled ocean-atmosphere feedback was originally proposed by Bjerknes. From an oceanographic point of view, the equatorial cold tongue is caused by easterly winds. Were the Earth climate symmetric about the equator, cross-equatorial wind would vanish, and the cold tongue would be much weaker and have a very different zonal structure than is observed today.

During non-El Niño conditions, the Walker circulation is seen at the surface as easterly trade winds that move water and air warmed by the sun toward the west. This also creates ocean upwelling off the coasts of Peru and Ecuador and brings nutrient-rich cold water to the surface, increasing fishing stocks. The western side of the equatorial Pacific is characterized by warm, wet, low-pressure weather as the collected moisture is dumped in the form of typhoons and thunderstorms. The ocean is some 60 cm (24 in) higher in the western Pacific as the result of this motion.

Sea Surface Temperature Oscillation

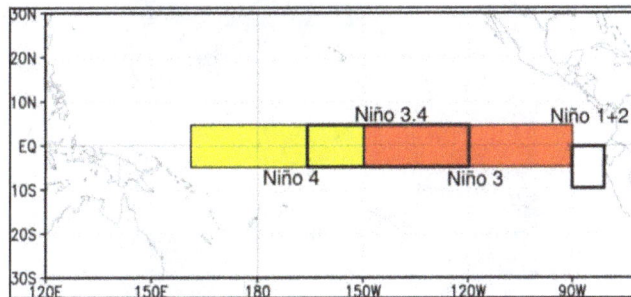

The various "Niño regions" where sea surface temperatures are monitored to determine the current ENSO phase (warm or cold)

Within the National Oceanic and Atmospheric Administration in the United States, sea surface temperatures in the Niño 3.4 region, which stretches from the 120th to 170th meridians west longitude astride the equator five degrees of latitude on either side, is monitored. This region is approximately 3,000 kilometres (1,900 mi) to the southeast of Hawaii. The most recent three-month average for the area is computed, and if the region is more than 0.5°C (0.9°F) above (or below) normal for that period, then an El Niño (or La Niña) is considered in progress. The United Kingdom's Met Office also uses a several month period to determine ENSO state. When this warming or cooling occurs for only seven to nine months, it is classified as El Niño/La Niña "conditions"; when it occurs for more than that period, it is classified as El Niño/La Niña "episodes".

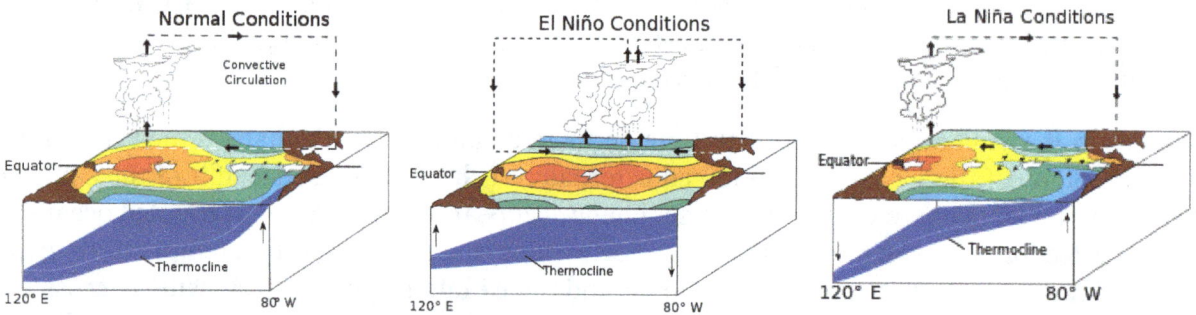

Normal Pacific pattern: Equatorial winds gather warm water pool toward the west. Cold water upwells along South American coast. (NOAA / PMEL / TAO).

El Niño conditions: Warm water pool approaches the South American coast. The absence of cold upwelling increases warming.

La Niña conditions: Warm water is farther west than usual.

Neutral Phase

If the temperature variation from climatology is within 0.5°C (0.9°F), ENSO conditions are described as neutral. Neutral conditions are the transition between warm and cold phases of ENSO. Ocean temperatures (by definition), tropical precipitation, and wind patterns are near average conditions during this phase. Close to half of all years are within neutral periods. During the neutral ENSO phase, other climate anomalies/patterns such as the sign of the North Atlantic Oscillation or the Pacific–North American teleconnection pattern exert more influence.

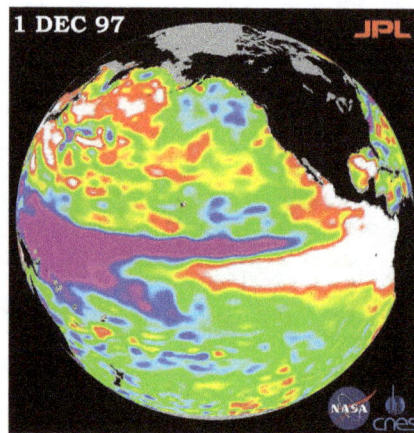

The 1997 El Niño observed by TOPEX/Poseidon

Warm Phase

When the Walker circulation weakens or reverses and the Hadley circulation strengthens an El Niño results, causing the ocean surface to be warmer than average, as upwelling of cold water occurs less or not at all offshore northwestern South America. El Niño is associated with a band of warmer than average ocean water temperatures that periodically develops off the Pacific coast of South America. *El niño* is Spanish for "the boy", and the capitalized term El Niño refers to the Christ child, Jesus, because periodic warming in the Pacific near South America is usually noticed around Christmas. It is a phase of 'El Niño–Southern Oscillation' (ENSO), which refers to variations in the temperature of the surface of the tropical eastern Pacific Ocean and in air surface pressure in the tropical western Pacific. The warm oceanic phase, El Niño, accompanies high air surface pressure in the western Pacific. Mechanisms that cause the oscillation remain under study.

Cold Phase

An especially strong Walker circulation causes a La Niña, resulting in cooler ocean temperatures in the central and eastern tropical Pacific Ocean due to increased upwelling. La Niña is a coupled ocean-atmosphere phenomenon that is the counterpart of El Niño as part of the broader El Niño Southern Oscillation climate pattern. The name La Niña originates from Spanish, meaning "the girl", analogous to El Niño meaning "the boy". During a period of La Niña, the sea surface temperature across the equatorial eastern central Pacific will be lower than normal by 3–5°C. In the United States, an *appearance* of La Niña happens for at least five months of La Niña conditions. However, each country and island nation has a different threshold for what constitutes a La Niña event, which is tailored to their specific interests. The Japan Meteorological Agency for example, declares that a La Niña event has started when the average five month sea surface temperature deviation for the NINO.3 region, is over 0.5°C (0.90°F) cooler for 6 consecutive months or longer.

Transitional Phases

Transitional phases at the onset or departure of El Niño or La Niña can also be important factors on global weather by affecting teleconnections. Significant episodes, known as Trans-Niño, are measured by the Trans-Niño index (TNI). Examples of affected short-time climate in North America include precipitation in the Northwest US and intense tornado activity in the contiguous US.

Southern Oscillation

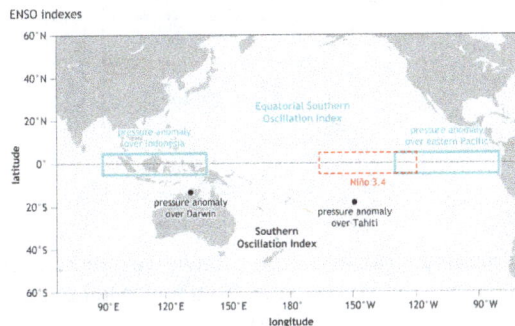

The regions where the air pressure are measured and compared to generate the Southern Oscillation Index.

The Southern Oscillation is the atmospheric component of El Niño. This component is an oscillation in surface air pressure between the tropical eastern and the western Pacific Ocean waters. The strength of the Southern Oscillation is measured by the Southern Oscillation Index (SOI). The SOI is computed from fluctuations in the surface air pressure difference between Tahiti (in the Pacific) and Darwin, Australia (on the Indian Ocean).

- El Niño episodes have negative SOI, meaning there is lower pressure over Tahiti and higher pressure in Darwin.

- La Niña episodes have positive SOI, meaning there is higher pressure in Tahiti and lower in Darwin.

Low atmospheric pressure tends to occur over warm water and high pressure occurs over cold water, in part because of deep convection over the warm water. El Niño episodes are defined as sustained warming of the central and eastern tropical Pacific Ocean, thus resulting in a decrease in the strength of the Pacific trade winds, and a reduction in rainfall over eastern and northern Australia. La Niña episodes are defined as sustained cooling of the central and eastern tropical Pacific Ocean, thus resulting in an increase in the strength of the Pacific trade winds, and the opposite effects in Australia when compared to El Niño.

Although the Southern Oscillation Index has a long station record going back to the 1800s, its reliability is limited due to the presence of both Darwin and Tahiti well south of the Equator, resulting in the surface air pressure at both locations being less directly related to ENSO. To overcome this question, a new index was created, being named Equatorial Southern Oscillation Index (EQSOI). To generate this index data, two new regions, centered on the Equator, were delimited to create a new index: The western one is located over Indonesia and the eastern one is located over equatorial Pacific, close to the South American coast. However, data on EQSOI goes back only to 1949.

Madden–Julian oscillation

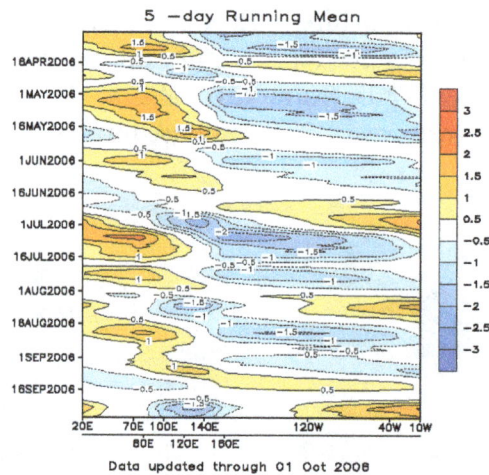

A Hovmöller diagram of the 5-day running mean of outgoing longwave radiation showing the MJO. Time increases from top to bottom in the figure, so contours that are oriented from upper-left to lower-right represent movement from west to east.

The Madden–Julian oscillation, or (MJO), is the largest element of the intraseasonal (30- to 90-day) variability in the tropical atmosphere, and was discovered by Roland Madden and Paul Julian of the National Center for Atmospheric Research (NCAR) in 1971. It is a large-scale coupling between atmospheric circulation and tropical deep convection. Rather than being a standing pattern like the El Niño Southern Oscillation (ENSO), the MJO is a traveling pattern that propagates eastward at approximately 4 to 8 m/s (9 to 18 mph), through the atmosphere above the warm parts of the Indian and Pacific oceans. This overall circulation pattern manifests itself in various ways, most clearly as anomalous rainfall. The wet phase of enhanced convection and precipitation is followed by a dry phase where thunderstorm activity is suppressed. Each cycle lasts approximately 30–60 days. Because of this pattern, The MJO is also known as the *30- to 60-day oscillation, 30- to 60-day wave,* or *intraseasonal oscillation.*

There is strong year-to-year (interannual) variability in MJO activity, with long periods of strong activity followed by periods in which the oscillation is weak or absent. This interannual variability of the MJO is partly linked to the El Niño–Southern Oscillation (ENSO) cycle. In the Pacific, strong MJO activity is often observed 6 – 12 months prior to the onset of an El Niño episode, but is virtually absent during the maxima of some El Niño episodes, while MJO activity is typically greater during a La Niña episode. Strong events in the Madden–Julian oscillation over a series of months in the western Pacific can speed the development of an El Niño or La Niña but usually do not in themselves lead to the onset of a warm or cold ENSO event. However, observations suggest that the 1982–1983 El Niño developed rapidly during July 1982 in direct response to a Kelvin wave triggered by an MJO event during late May. Further, changes in the structure of the MJO with the seasonal cycle and ENSO might facilitate more substantial impacts of the MJO on ENSO. For example, the surface westerly winds associated with active MJO convection are stronger during advancement toward El Niño and the surface easterly winds associated with the suppressed convective phase are stronger during advancement toward La Nina.

Impacts

On Precipitation

COLD EPISODE RELATIONSHIPS   DECEMBER - FEBRUARY

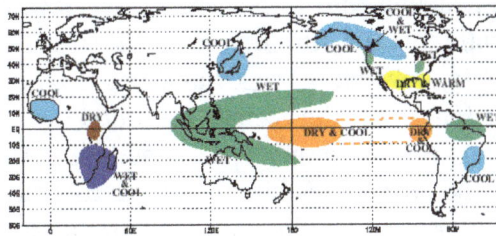

COLD EPISODE RELATIONSHIPS   JUNE - AUGUST

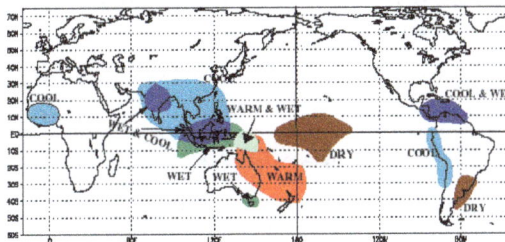

Regional impacts of La Niña.

Developing countries dependent upon agriculture and fishing, particularly those bordering the Pacific Ocean, are the most affected by ENSO. The effects of El Niño in South America are direct and strong. An El Niño is associated with warm and very wet weather months in April–October along the coasts of northern Peru and Ecuador, causing major flooding whenever the event is strong or extreme. La Niña causes a drop in sea surface temperatures over Southeast Asia and heavy rains over Malaysia, the Philippines, and Indonesia.

To the north across Alaska, La Niña events lead to drier than normal conditions, while El Niño events do not have a correlation towards dry or wet conditions. During El Niño events, increased precipitation is expected in California due to a more southerly, zonal, storm track. During La Niña,

increased precipitation is diverted into the Pacific Northwest due to a more northerly storm track. During La Niña events, the storm track shifts far enough northward to bring wetter than normal winter conditions (in the form of increased snowfall) to the Midwestern states, as well as hot and dry summers. During the El Niño portion of ENSO, increased precipitation falls along the Gulf coast and Southeast due to a stronger than normal, and more southerly, polar jet stream. In the late winter and spring during El Niño events, drier than average conditions can be expected in Hawaii. On Guam during El Niño years, dry season precipitation averages below normal. However, the threat of a tropical cyclone is over triple what is normal during El Niño years, so extreme shorter duration rainfall events are possible. On American Samoa during El Niño events, precipitation averages about 10 percent above normal, while La Niña events lead to precipitation amounts which average close to 10 percent below normal. ENSO is linked to rainfall over Puerto Rico. During an El Niño, snowfall is greater than average across the southern Rockies and Sierra Nevada mountain range, and is well-below normal across the Upper Midwest and Great Lakes states. During a La Niña, snowfall is above normal across the Pacific Northwest and western Great Lakes.

On Tehuantepecers

The synoptic condition for the Tehuantepecer, a violent mountain-gap wind in between the mountains of Mexico and Guatemala, is associated with high-pressure system forming in Sierra Madre of Mexico in the wake of an advancing cold front, which causes winds to accelerate through the Isthmus of Tehuantepec. Tehuantepecers primarily occur during the cold season months for the region in the wake of cold fronts, between October and February, with a summer maximum in July caused by the westward extension of the Azores-Bermuda high pressure system. Wind magnitude is greater during El Niño years than during La Niña years, due to the more frequent cold frontal incursions during El Niño winters. Tehuantepec winds reach 20 knots (40 km/h) to 45 knots (80 km/h), and on rare occasions 100 knots (200 km/h). The wind's direction is from the north to north-northeast. It leads to a localized acceleration of the trade winds in the region, and can enhance thunderstorm activity when it interacts with the Intertropical Convergence Zone. The effects can last from a few hours to six days.

On Global Warming

NOAA graph of Global Annual Temperature Anomalies 1950–2012, showing ENSO

El Niño events cause short-term (approximately 1 year in length) spikes in global average surface temperature while La Niña events cause short term cooling. Therefore, the relative frequency of El

Niño compared to La Niña events can affect global temperature trends on decadal timescales. Over the last several decades, the number of El Niño events increased, and the number of La Niña events decreased, although observation of ENSO for much longer is needed to detect robust changes.

The studies of historical data show the recent El Niño variation is most likely linked to global warming. For example, one of the most recent results, even after subtracting the positive influence of decadal variation, is shown to be possibly present in the ENSO trend, the amplitude of the ENSO variability in the observed data still increases, by as much as 60% in the last 50 years.

The exact changes happening to ENSO in the future is uncertain: Different models make different predictions. It may be that the observed phenomenon of more frequent and stronger El Niño events occurs only in the initial phase of the global warming, and then (e.g., after the lower layers of the ocean get warmer, as well), El Niño will become weaker than it was. It may also be that the stabilizing and destabilizing forces influencing the phenomenon will eventually compensate for each other. More research is needed to provide a better answer to that question. The ENSO is considered to be a potential tipping element in Earth's climate and, under the global warming, can enhance or alternate regional climate extreme events through a strengthened teleconnection. For example, an increase in the frequency and magnitude of El Niño events have triggered warmer than usual temperatures over the Indian Ocean, by modulating the Walker circulation. This has resulted in a rapid warming of the Indian Ocean, and consequently a weakening of the Asian Monsoon.

On Coral Bleaching

Following the El Nino event in 1997 – 1998, the largest recorded to date, the Pacific Marine Environmental Laboratory attributes the first large-scale water bleaching event to the warming waters. This process, called reef symbiosis, results from warming ocean water temperatures.

Diversity

The traditional ENSO (El Niño Southern Oscillation), also called Eastern Pacific (EP) ENSO, involves temperature anomalies in the eastern pacific. However, in the 1990s and 2000s, nontraditional ENSO conditions were observed, in which the usual place of the temperature anomaly (Niño 1 and 2) is not affected, but an anomaly arises in the central Pacific (Niño 3.4). The phenomenon is called Central Pacific (CP) ENSO, "dateline" ENSO (because the anomaly arises near the dateline), or ENSO "Modoki" (Modoki is Japanese for "similar, but different"). There are flavors of ENSO additional to EP and CP types and some scientists argue that ENSO exists as a continuum often with hybrid types.

The effects of the CP ENSO are different from those of the traditional EP ENSO. The El Niño Modoki leads to more hurricanes more frequently making landfall in the Atlantic. La Niña Modoki leads to a rainfall increase over northwestern Australia and northern Murray–Darling basin, rather than over the east as in a conventional La Niña. Also, La Niña Modoki increases the frequency of cyclonic storms over Bay of Bengal, but decreases the occurrence of severe storms in the Indian Ocean.

The recent discovery of ENSO Modoki has some scientists believing it to be linked to global warming. However, comprehensive satellite data go back only to 1979. More research must be done to find the correlation and study past El Niño episodes. More generally, there is no scientific consensus on how/if climate change may affect ENSO.

There is also a scientific debate on the very existence of this "new" ENSO. Indeed, a number of studies dispute the reality of this statistical distinction or its increasing occurrence, or both, either arguing the reliable record is too short to detect such a distinction, finding no distinction or trend using other statistical approaches, or that other types should be distinguished, such as standard and extreme ENSO. Following the asymmetric nature of the warm and cold phases of ENSO, some studies could not identify such distinctions for La Niña, both in observations and in the climate models, but some sources indicate that there is a variation on La Niña with cooler waters on central Pacific and average or warmer water temperatures on both eastern and western Pacific, also showing eastern Pacific Ocean currents going to the opposite direction compared to the currents in traditional La Niñas.

References

- Junling Huang and Michael B. McElroy (2014). "Contributions of the Hadley and Ferrel Circulations to the Energetics of the Atmosphere over the Past 32 Years". Journal of Climate. 27 (7): 2656–2666. Bibcode:2014J-Cli...27.2656H. doi:10.1175/jcli-d-13-00538.1

- Halldór Björnsson. "Global circulation". Archived from the original on March 24, 2010. Retrieved September 2, 2016. . Veðurstofa Íslands. Retrieved on 2008-06-15

- Robock, Alan (2000). "Volcanic eruptions and climate". Reviews of Geophysics. 38 (2): 191–219. Bibcode:2000RvGeo..38..191R. doi:10.1029/1998RG000054

- Erik A. Rasmussen and John Turner (2003). Polar lows: mesoscale weather systems in the polar regions. Cambridge University Press. p. 174. ISBN 978-0-521-62430-5

- Hartmann, D; Schoeberl, M (1991). "Mixing of polar vortex air into middle latitudes as revealed by tracer-tracer scatterplots". Journal of Geophysical Research. 102: 13119. Bibcode:1997JGR...10213119W. doi:10.1029/96JD03715

- Todd Mitchell (2004). Arctic Oscillation (AO) time series, 1899 – June 2002. University of Washington. Retrieved on 2009-03-02

- Reichler, Tom; Kim, J; Manzini, E; Kroger, J (2012). "A stratospheric connection to Atlantic climate variability". Nature Geoscience. 5: 783–787. Bibcode:2012NatGe...5..783R. doi:10.1038/ngeo1586

- Rolf Müller (2010). Tracer-tracer Relations as a Tool for Research on Polar Ozone Loss. Forschungszentrum Jülich. p. 47. ISBN 978-3-89336-614-9

- Cavallo, S; Hakim, G.J. (2013). "Physical mechanisms of tropopause polar vortex intensity change". Journal of the Atmospheric Sciences. 70 (11): 3359–3373. doi:10.1175/JAS-D-13-088.1

- "GEOS-5 Analyses and Forecasts of the Major Stratospheric Sudden Warming of January 2013" (Press release). Goddard Space Flight Center. Retrieved January 8, 2014

- Reichler, Tom; Kim, J; Manzini, E; Kroger, J (2012). "A stratospheric connection to Atlantic climate variability". Nature Geoscience. 5: 783–787. Bibcode:2012NatGe...5..783R. doi:10.1038/ngeo1586

- K. Mohanakuma (2008). Stratosphere troposphere interactions: an introduction. Springer. p. 34. ISBN 978-1-4020-8216-0

- Climate Prediction Center (2005-12-19). "Frequently Asked Questions about El Niño and La Niña". National Centers for Environmental Prediction. Retrieved 2009-07-17

- Screen, J A (2013). "Influence of Arctic sea ice on European summer precipitation". Environmental Research Letters. 8 (4): 044015. Bibcode:2013ERL.....8d4015S. doi:10.1088/1748-9326/8/4/044015

- Manney, G; Zurek, R; O'Neill, A; Swinbank, R (1994). "On the motion of air through the stratospheric polar vortex". Journal of the Atmospheric Sciences. 51 (20): 2973–2994. Bibcode:1994JAtS...51.2973M. doi:10.1175 /1520-0469(1994)051<2973:otmoat>2.0.co;2

- Jennings, S., Kaiser, M.J., Reynolds, J.D. (2001) "Marine Fisheries Ecology." Oxford: Blackwell Science Ltd. ISBN 0-632-05098-5

- Masato, Giacomo; Hoskins, Brian J.; Woollings, Tim (2013). "Winter and Summer Northern Hemisphere Blocking in CMIP5 Models". Journal of Climate. 26 (18): 7044–59. doi:10.1175/JCLI-D-12-00466.1

- "Envisat watches for La Niña". BNSC via the Internet Wayback Machine. 2011-01-09. Archived from the original on 2008-04-24. Retrieved 2007-07-26

- Trenberth, Kevin E.; D. P. Stepaniak (2001). "Indices of El Niño Evolution". J. Climate. 14 (8): 1697–701. Bibcode:2001JCli...14.1697T. ISSN 1520-0442. doi:10.1175/1520-0442(2001)014<1697:LIOENO>2.0.CO;2

- Brown, Patrick T.; Li, Wenhong; Xie, Shang-Ping (2015-01-27). "Regions of significant influence on unforced global mean surface air temperature variability in climate models". Journal of Geophysical Research: Atmospheres. 120 (2): 2014JD022576. Bibcode:2015JGRD..120..480B. ISSN 2169-8996. doi:10.1002/ 2014JD022576

- International Research Institute for Climate and Society (February 2002). "More Technical ENSO Comment". Columbia University. Retrieved 2014-06-30

- Roundy, Paul E.; Kiladis, George N. (2007). "Analysis of a Reconstructed Oceanic Kelvin Wave Dynamic Height Dataset for the Period 1974–2005". ametsoc.org. p. 4341. doi:10.1175/JCLI4249.1. Retrieved 20 October 2015

6

Atmospheric Thermodynamics

The thermodynamic effects that occur in the Earth's atmosphere and causes atmospheric phenomena such as weather and climate are studied under atmospheric thermodynamics. Related topics such as atmospheric convection, atmospheric instability, atmospheric sounding have been included in this section. The aspects elucidated in this chapter are of vital importance, and provide a better understanding of the Earth's atmosphere.

Atmospheric Thermodynamics

Atmospheric thermodynamics is the study of heat-to-work transformations (and their reverse) that take place in the earth's atmosphere and manifest as weather or climate. Atmospheric thermodynamics use the laws of classical thermodynamics, to describe and explain such phenomena as the properties of moist air, the formation of clouds, atmospheric convection, boundary layer meteorology, and vertical instabilities in the atmosphere. Atmospheric thermodynamic diagrams are used as tools in the forecasting of storm development. Atmospheric thermodynamics forms a basis for cloud microphysics and convection parameterizations used in numerical weather models and is used in many climate considerations, including convective-equilibrium climate models.

Overview

The atmosphere is an example of a non-equilibrium system. Atmospheric thermodynamics describes the effect of buoyant forces that cause the rise of less dense (warmer) air, the descent of more dense air , and the transformation of water from liquid to vapor (evaporation) and its condensation. Those dynamics are modified by the force of the pressure gradient and that motion is modified by the Coriolis force. The tools used include the law of energy conservation, the ideal gas law, specific heat capacities, the assumption of isentropic processes (in which entropy is a constant), and moist adiabatic processes (during which no energy is transferred as heat). Most of tropospheric gases are treated as ideal gases and water vapor, with its ability to change phase from vapor, to liquid, to solid, and back is considered as one of the most important trace components of air.

Advanced topics are phase transitions of water, homogeneous and in-homogeneous nucleation, effect of dissolved substances on cloud condensation, role of supersaturation on formation of ice crystals and cloud droplets. Considerations of moist air and cloud theories typically involve various temperatures, such as equivalent potential temperature, wet-bulb and virtual temperatures. Connected areas are energy, momentum, and mass transfer, turbulence interaction between air particles in clouds, convection, dynamics of tropical cyclones, and large scale dynamics of the atmosphere.

The major role of atmospheric thermodynamics is expressed in terms of adiabatic and diabatic forces acting on air parcels included in primitive equations of air motion either as grid resolved or subgrid parameterizations. These equations form a basis for the numerical weather and climate predictions.

History

In the early 19th century thermodynamicists such as Sadi Carnot, Rudolf Clausius, and Émile Clapeyron developed mathematical models on the dynamics of fluid bodies and vapors related to the combustion and pressure cycles of atmospheric steam engines; one example is the Clausius–Clapeyron equation. In 1873, thermodynamicist Willard Gibbs published "Graphical Methods in the Thermodynamics of Fluids."

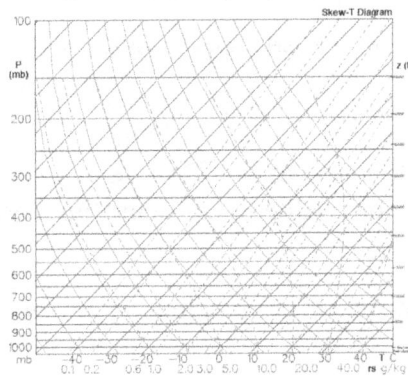

Thermodynamic diagram developed in the 19th century is still used to calculate quantities such as convective available potential energy or air stability.

These sorts of foundations naturally began to be applied towards the development of theoretical models of atmospheric thermodynamics which drew the attention of the best minds. Papers on atmospheric thermodynamics appeared in the 1860s that treated such topics as dry and moist adiabatic processes. In 1884 Heinrich Hertz devised first atmospheric thermodynamic diagram (emagram). Pseudo-adiabatic process was coined by von Bezold describing air as it is lifted, expands, cools, and eventually precipitates its water vapor; in 1888 he published voluminous work entitled "On the thermodynamics of the atmosphere".

In 1911 von Alfred Wegener published a book "Thermodynamik der Atmosphäre", Leipzig, J. A. Barth. From here the development of atmospheric thermodynamics as a branch of science began to take root. The term "atmospheric thermodynamics", itself, can be traced to Frank W. Verys 1919 publication: "The radiant properties of the earth from the standpoint of atmospheric thermodynamics" (Occasional scientific papers of the Westwood Astrophysical Observatory). By the late 1970s various textbooks on the subject began to appear. Today, atmospheric thermodynamics is an integral part of weather forecasting.

Chronology

- 1751 Charles Le Roy recognized dew point temperature as point of saturation of air

- 1782 Jacques Charles made hydrogen balloon flight measuring temperature and pressure in Paris

- 1784 Concept of variation of temperature with height was suggested

- 1801-1803 John Dalton developed his laws of pressures of vapours

- 1804 Joseph Louis Gay-Lussac made balloon ascent to study weather

- 1805 Pierre Simon Laplace developed his law of pressure variation with height

- 1841 James Pollard Espy publishes paper on convection theory of cyclone energy

- 1889 Hermann von Helmholtz and John William von Bezold used the concept of potential temperature, von Bezold used adiabatic lapse rate and pseudoadiabat

- 1893 Richard Asman constructs first aerological sonde (pressure-temperature-humidity)

- 1894 John Wilhelm von Bezold used concept of equivalent temperature

- 1926 Sir Napier Shaw introduced tephigram

- 1933 Tor Bergeron published paper on "Physics of Clouds and Precipitation" describing precipitation from supercooled (due to condensational growth of ice crystals in presence of water drops)

- 1946 Vincent J. Schaeffer and Irving Langmuir performed the first cloud-seeding experiment

- 1986 K. Emanuel conceptualizes tropical cyclone as Carnot heat engine

Applications

Hadley Circulation

The Hadley Circulation can be considered as a heat engine. The Hadley circulation is identified with rising of warm and moist air in the equatorial region with the descent of colder air in the sub-tropics corresponding to a thermally driven direct circulation, with consequent net production of kinetic energy. The thermodynamic efficiency of the Hadley system, considered as a heat engine, has been relatively constant over the 1979~2010 period, averaging 2.6%. Over the same interval, the power generated by the Hadley regime has risen at an average rate of about 0.54 TW per yr; this reflects an increase in energy input to the system consistent with the observed trend in the tropical sea surface temperatures.

Tropical Cyclone Carnot Cycle

The thermodynamic behavior of a hurricane can be modelled as a heat engine that operates between the heat reservoir of the sea at a temperature of about 300K (27°C) and the heat sink of the tropopause at a temperature of about 200K (-72°C) and in the process converts heat energy into mechanical energy of winds. Parcels of air traveling close to the sea surface take up heat and water vapor, the warmed air rises and expands and cools as it does so causes condensation and precipitation. The rising air, and condensation, produces circulatory winds that are propelled by the Coriolis force, which whip up waves and increase the amount of warm moist air that powers the cyclone. Both a decreasing temperature in the upper troposphere or an increasing temperature of the atmosphere close to the surface will increase the maximum winds observed in hurricanes. When applied to hurricane dynamics it defines a Carnot heat engine cycle and predicts maximum hurricane intensity.

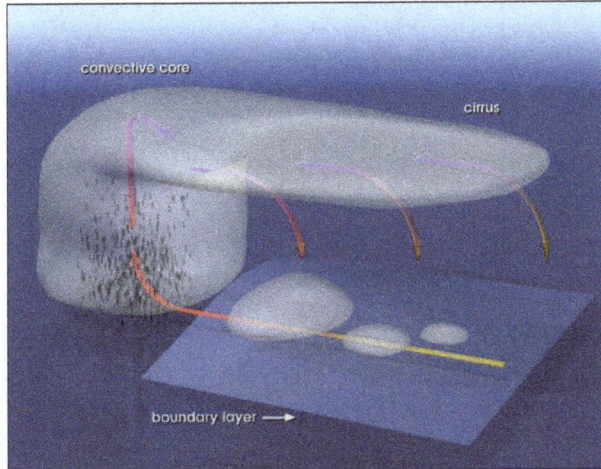

Air is being moistened as it travels toward convective system. Ascending motion in a deep convective core produces air expansion, cooling, and condensation. Upper level outflow visible as an anvil cloud is eventually descending conserving mass (rysunek - Robert Simmon).

Water Vapor and Global Climate Change

The Clausius–Clapeyron relation shows how the water-holding capacity of the atmosphere increases by about 8% per Celsius increase in temperature. (It does not directly depend on other parameters like the pressure or density.) This water-holding capacity, or "equilibrium vapor pressure", can be approximated using the August-Roche-Magnus formula

$$e_s(T) = 6.1094 \exp\left(\frac{17.625T}{T + 243.04}\right)$$

(where $e_s(T)$ is the equilibrium or saturation vapor pressure in hPa, and T is temperature in degrees Celsius). This shows that when atmospheric temperature increases (e.g., due to greenhouse gases) the absolute humidity should also increase exponentially (assuming a constant relative humidity). However, this purely thermodynamic argument is subject of considerable debate because convective processes might cause extensive drying due to increased areas of subsidence, efficiency of precipitation could be influenced by the intensity of convection, and because cloud formation is related to relative humidity.

Atmospheric Convection

Atmospheric convection is the result of a parcel-environment instability, or temperature difference, layer in the atmosphere. Different lapse rates within dry and moist air lead to instability. Mixing of air during the day which expands the height of the planetary boundary layer leads to increased winds, cumulus cloud development, and decreased surface dew points. Moist convection leads to thunderstorm development, which is often responsible for severe weather throughout the world. Special threats from thunderstorms include hail, downbursts, and tornadoes.

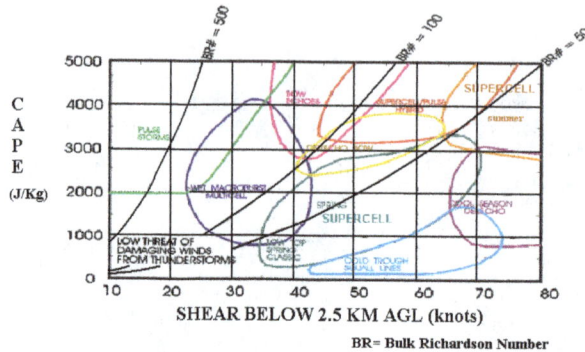

Conditions favorable for thunderstorm types and complexes

Overview

There are a few general archetypes of atmospheric instability that are used to explain convection (or lack thereof). A necessary (but not sufficient) condition for convection is that the environmental lapse rate (the rate of decrease of temperature with height) is steeper than the lapse rate experienced by a rising parcel of air. When this condition is met, upward-displaced air parcels can become buoyant and thus experience a further upward force. Buoyant convection begins at the level of free convection (LFC), above which an air parcel may ascend through the free convective layer (FCL) with positive buoyancy. Its buoyancy turns negative at the equilibrium level (EL), but the parcel's vertical momentum may carry it to the maximum parcel level (MPL) where the negative buoyancy decelerates the parcel to a stop. Integrating the buoyancy force over the parcel's vertical displacement yields Convective Available Potential Energy (CAPE), the Joules of energy available per kilogram of potentially buoyant air. CAPE is an upper limit for an ideal undilute parcel, and the square root of twice the CAPE is sometimes called a thermodynamic speed limit for updrafts, based on the simple kinetic energy equation.

However, such buoyant Acceleration concepts give an oversimplified view of convection. Drag is an opposite force to counter buoyancy , so that parcel ascent occurs under a balance of forces, like the terminal velocity of a falling object. Buoyancy may be reduced by entrainment, which dilutes the parcel with environmental air.

Atmospheric convection is called *deep* when it extends from near the surface to above the 500 hPa level, generally stopping at the tropopause at around 200 hPa. Most atmospheric deep convection occurs in the tropics as the rising branch of the Hadley circulation; and represents a strong local coupling between the surface and the upper troposphere which is largely absent in winter midlatitudes. Its counterpart in the ocean (deep convection downward in the water column) only occurs at a few locations. While less dynamically important than in the atmosphere, such oceanic convection is responsible for the worldwide existence of cold water in the lowest layers of the ocean.

Initiation

A thermal column (or thermal) is a vertical section of rising air in the lower altitudes of the Earth's atmosphere. Thermals are created by the uneven heating of the Earth's surface from solar radiation. The Sun warms the ground, which in turn warms the air directly above it. The warmer air expands, becoming less dense than the surrounding air mass, and creating a thermal low. The mass of lighter air rises, and as it does, it cools due to its expansion at lower high-altitude pressures. It stops rising when it has cooled to the same temperature as the surrounding air. Associated with a thermal is a downward flow surrounding the thermal column. The downward moving exterior is caused by colder air being displaced at the top of the thermal. Another convection-driven weather effect is the sea breeze.

Thunderstorms

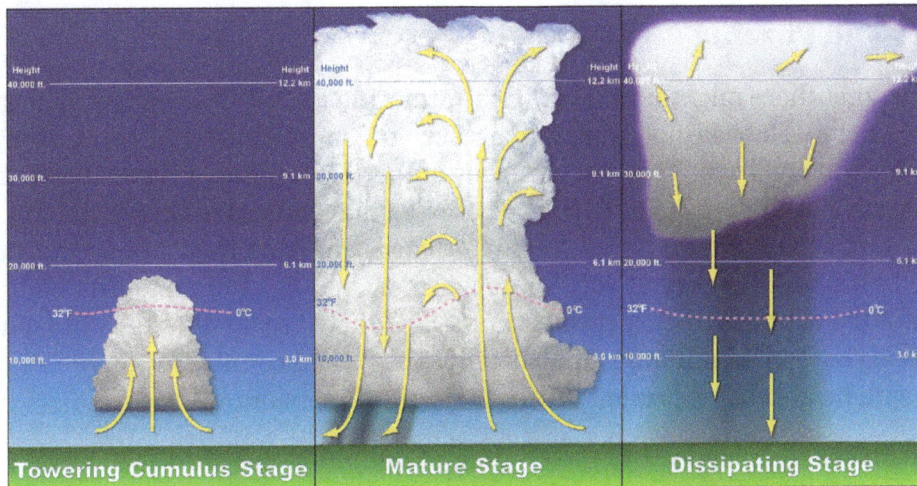

Stages of a thunderstorm's life.

Warm air has a lower density than cool air, so warm air rises within cooler air, similar to hot air balloons. Clouds form as relatively warmer air carrying moisture rises within cooler air. As the moist air rises, it cools causing some of the water vapor in the rising packet of air to condense. When the moisture condenses, it releases energy known as latent heat of vaporization which allows the rising packet of air to cool less than its surrounding air, continuing the cloud's ascension. If enough instability is present in the atmosphere, this process will continue long enough for cumulonimbus clouds to form, which support lightning and thunder. Generally, thunderstorms require three conditions to form: moisture, an unstable airmass, and a lifting force (heat).

All thunderstorms, regardless of type, go through three stages: the developing stage, the mature stage, and the dissipation stage. The average thunderstorm has a 24 km (15 mi) diameter. Depending on the conditions present in the atmosphere, these three stages take an average of 30 minutes to go through.

There are four main types of thunderstorms: single-cell, multicell, squall line (also called multicell line) and supercell. Which type forms depends on the instability and relative wind conditions at different layers of the atmosphere ("wind shear"). Single-cell thunderstorms form in environments of low vertical wind shear and last only 20–30 minutes. Organized thunderstorms and thunderstorm clusters/lines can have longer life cycles as they form in environments of significant vertical wind shear, which aids the development of stronger updrafts as well as various forms of severe weather. The supercell is the strongest of the thunderstorms, most commonly associated with large hail, high winds, and tornado formation.

The latent heat release from condensation is the determinate between significant convection and almost no convection at all. The fact that air is generally cooler during winter months, and therefore cannot hold as much water vapor and associated latent heat, is why significant convection (thunderstorms) are infrequent in cooler areas during that period. Thundersnow is one situation where forcing mechanisms provide support for very steep environmental lapse rates, which as mentioned before is an archetype for favored convection. The small amount of latent heat released from air rising and condensing moisture in a thundersnow also serves to increase this convective potential, although minimally. There are also three types of thunderstorms: orographic, air mass, and frontal.

Boundaries and Forcing

Despite the fact that there might be a layer in the atmosphere that has positive values of CAPE, if the parcel does not reach or begin rising to that level, the most significant convection that occurs in the FCL will not be realized. This can occur for numerous reasons. Primarily, it is the result of a cap, or convective inhibition (CIN/CINH). Processes that can erode this inhibition are heating of the Earth's surface and forcing. Such forcing mechanisms encourage upward vertical velocity, characterized by a speed that is relatively low to what you find in a thunderstorm updraft. Because of this, it is not the actual air being pushed to its LFC that "breaks through" the inhibition, but rather the forcing cools the inhibition adiabatically. This would counter, or "erode" the increase of temperature with height that is present during a capping inversion.

Forcing mechanisms that can lead to the eroding of inhibition are ones that create some sort of evacuation of mass in the upper parts of the atmosphere, or a surplus of mass in the low levels of the atmosphere, which would lead to upper level divergence or lower level convergence, respectively. Upward vertical motion will often follow. Specifically, a cold front, sea/lake breeze, outflow boundary, or forcing through vorticity dynamics (differential positive vorticity advection) of the atmosphere such as with troughs, both shortwave and longwave. Jet streak dynamics through the imbalance of Coriolis and pressure gradient forces, causing subgeostrophic and supergeostrophic flows, can also create upward vertical velocities. There are numerous other atmospheric setups in which upward vertical velocities can be created.

Concerns Regarding Severe Deep Moist Convection

Buoyancy is key to thunderstorm growth and is necessary for any of the severe threats within a thunderstorm. There are other processes, not necessarily thermodynamic, that can increase updraft strength. These include updraft rotation, low level convergence, and evacuation of mass out of the top of the updraft via strong upper level winds and the jet stream.

Hail

Like other precipitation in cumulonimbus clouds hail begins as water droplets. As the droplets rise and the temperature goes below freezing, they become supercooled water and will freeze on contact with condensation nuclei. A cross-section through a large hailstone shows an onion-like structure. This means the hailstone is made of thick and translucent layers, alternating with layers that are thin, white and opaque. Former theory suggested that hailstones were subjected to multiple descents and ascents, falling into a zone of humidity and refreezing as they were uplifted. This up and down motion was thought to be responsible for the successive layers of the hailstone. New research (based on theory and field study) has shown this is not necessarily true.

Hail shaft

Severe thunderstorms containing hail can exhibit a characteristic green coloration.

The storm's updraft, with upwardly directed wind speeds as high as 180 kilometres per hour (110 mph), blow the forming hailstones up the cloud. As the hailstone ascends it passes into areas of the cloud where the concentration of humidity and supercooled water droplets varies. The hailstone's growth rate changes depending on the variation in humidity and supercooled water droplets that it encounters. The accretion rate of these water droplets is another factor in the hailstone's growth. When the hailstone moves into an area with a high concentration of water droplets, it captures the latter and acquires a translucent layer. Should the hailstone move into an area where mostly water vapour is available, it acquires a layer of opaque white ice.

Furthermore, the hailstone's speed depends on its position in the cloud's updraft and its mass. This determines the varying thicknesses of the layers of the hailstone. The accretion rate of supercooled water droplets onto the hailstone depends on the relative velocities between these water droplets and the hailstone itself. This means that generally the larger hailstones will form some distance from the stronger updraft where they can pass more time growing As the hailstone grows it releases latent heat, which keeps its exterior in a liquid phase. Undergoing 'wet growth', the outer layer is *sticky,* or more adhesive, so a single hailstone may grow by collision with other smaller hailstones, forming a larger entity with an irregular shape.

The hailstone will keep rising in the thunderstorm until its mass can no longer be supported by the updraft. This may take at least 30 minutes based on the force of the updrafts in the hail-producing thunderstorm, whose top is usually greater than 10 kilometres (6.2 mi) high. It then falls toward the ground while continuing to grow, based on the same processes, until it leaves the cloud. It will later begin to melt as it passes into air above freezing temperature.

Thus, a unique trajectory in the thunderstorm is sufficient to explain the layer-like structure of the hailstone. The only case in which we can discuss multiple trajectories is in a multicellular thunderstorm where the hailstone may be ejected from the top of the "mother" cell and captured in the updraft of a more intense "daughter cell". This however is an exceptional case.

Downburst

A downburst is created by a column of sinking air that, after hitting ground level, spreads out in all directions and is capable of producing damaging straight-line winds of over 240 kilometres per hour (150 mph), often producing damage similar to, but distinguishable from, that caused by tornadoes. This is because the physical properties of a downburst are completely different from those of a tornado. Downburst damage will radiate from a central point as the descending column spreads out when impacting the surface, whereas tornado damage tends towards convergent damage consistent with rotating winds. To differentiate between tornado damage and damage from a downburst, the term straight-line winds is applied to damage from microbursts.

Downbursts are particularly strong downdrafts from thunderstorms. Downbursts in air that is precipitation free or contains virga are known as dry downbursts; those accompanied with precipitation are known as wet downbursts. Most downbursts are less than 4 kilometres (2.5 mi) in extent: these are called microbursts. Downbursts larger than 4 kilometres (2.5 mi) in extent are sometimes called macrobursts. Downbursts can occur over large areas. In the extreme case, a derecho can cover a huge area more than 320 kilometres (200 mi) wide and over 1,600 kilometres (990 mi) long, lasting up to 12 hours or more, and is associated with some of the most intense straight-line winds, but the generative process is somewhat different from that of most downbursts.

Tornadoes

The F5 tornado that struck Elie, Manitoba in 2007

A tornado is a dangerous rotating column of air in contact with both the surface of the earth and the base of a cumulonimbus cloud (thundercloud) or a cumulus cloud, in rare cases. Tornadoes come in many sizes but typically form a visible condensation funnel whose narrowest end reaches the earth and surrounded by a cloud of debris and dust.

Tornadoes wind speeds generally average between 64 kilometres per hour (40 mph) and 180 kilometres per hour (110 mph). They are approximately 75 metres (246 ft) across and travel a few kilometers before dissipating. Some attain wind speeds in excess of 480 kilometres per hour (300 mph), may stretch more than a 1.6 kilometres (0.99 mi) across, and maintain contact with the ground for more than 100 kilometres (62 mi).

Tornadoes, despite being one of the most destructive weather phenomena are generally short-lived. A long-lived tornado generally lasts no more than an hour, but some have been known to last for 2 hours or longer (for example, the Tri-state tornado). Due to their relatively short duration, less information is known about the development and formation of tornadoes.

Measurement

The potential for convection in the atmosphere is often measured by an atmospheric temperature/dewpoint profile with height. This is often displayed on a Skew-T chart or other similar thermodynamic diagram. These can be plotted by a measured sounding analysis, which is the sending of a radiosonde attached to a balloon into the atmosphere to take the measurements with height. Forecast models can also create these diagrams, but are less accurate due to model uncertainties and biases, and have lower spatial resolution. Although, the temporal resolution of forecast model soundings is greater than the direct measurements, where the former can have plots for intervals of up to every 3 hours, and the latter as having only 2 per day (although when a convective event is expected a special sounding might be taken outside of the normal schedule of 00Z and then 12Z.).

Other Forecasting Concerns

Atmospheric convection can also be responsible for and have implications on a number of other weather conditions. A few examples on the smaller scale would include: Convection mixing the planetary boundary layer (PBL) and allowing drier air aloft to the surface thereby decreasing dew points, creating cumulus-type clouds which can limit a small amount of sunshine, increasing surface winds, making outflow boundaries/and other smaller boundaries more diffuse, and the eastward propagation of the dryline during the day. On the larger scale, rising of air can lead to warm core surface lows, often found in the desert southwest.

Atmospheric Instability

Atmospheric instability is a condition where the Earth's atmosphere is generally considered to be unstable and as a result the weather is subjected to a high degree of variability through distance and time. Atmospheric stability is a measure of the atmosphere's tendency to encourage or deter vertical motion, and vertical motion is directly correlated to different types of weather systems and their severity. In unstable conditions, a lifted thing, such as a parcel of air will be warmer than the surrounding air at altitude. Because it is warmer, it is less dense and is prone to further ascent.

A dust devil in Ramadi, Iraq

In meteorology, instability can be described by various indices such as the Bulk Richardson Number, lifted index, K-index, convective available potential energy (CAPE), the Showalter, and the Vertical totals. These indices, as well as atmospheric instability itself, involve temperature changes through the troposphere with height, or lapse rate. Effects of atmospheric instability in moist atmospheres include thunderstorm development, which over warm oceans can lead to tropical cyclogenesis, and turbulence. In dry atmospheres, inferior mirages, dust devils, steam devils, and fire whirls can form. Stable atmospheres can be associated with drizzle, fog, increased air pollution, a lack of turbulence, and undular bore formation.

Forms

Anvil shaped thundercloud in the mature stage over Swifts Creek, Victoria

There are two primary forms of atmospheric instability:

- Convective instability
- Dynamic instability (fluid mechanics)

Under convective instability thermal mixing through convection in the form of warm air rising leads to the development of clouds and possibly precipitation or convective storms. Dynamic instability is produced through the horizontal movement of air and the physical forces it is subjected to such as the Coriolis force and pressure gradient force. Dynamic lifting and mixing produces cloud, precipitation and storms often on a synoptic scale.

Cause of Instability

Whether or not the atmosphere has stability depends partially on the moisture content. In a very dry troposphere, a temperature decrease with height less than 9.8C per kilometer ascent indicates stability, while greater changes indicate instability. This lapse rate is known as the dry adiabatic lapse rate. In a completely moist troposphere, a temperature decrease with height less than 6C per kilometer ascent indicates stability, while greater changes indicate instability. In the range between 6C and 9.8C temperature decrease per kilometer ascent, the term conditionally unstable is used.

Indices used for its Determination

Lifted Index

The lifted index (LI), usually expressed in Kelvins, is the temperature difference between an air parcel lifted adiabatically $T_p(p)$ and the temperature of the environment $T_e(p)$ at a given pressure height in the troposphere, usually 500 hPa (mb). When the value is positive, the atmosphere (at the respective height) is stable and when the value is negative, the atmosphere is unstable. Thunderstorms are expected with values below -2, and severe weather is anticipated with values below -6.

K Index

K-index value	Thunderstorm Probability
Less than 20	None
20 to 25	Isolated thunderstorms
26 to 30	Widely scattered thunderstorms
31 to 35	Scattered thunderstorms
Above 35	Numerous thunderstorms

The K index is derived arithmetically: K-index = (850 hPa temperature - 500 hPa temperature) + 850 hPa dew point - 700 hPa dew point depression:

- The temperature difference between 850 hPa (5,000 feet (1,500 m) above sea level) and 500 hPa (18,000 feet (5,500 m) above sea level) is used to parameterize the vertical temperature lapse rate.

- The 850 hPa dew point provides information on the moisture content of the lower atmosphere.

- The vertical extent of the moist layer is represented by the difference of the 700 hPa temperature (10,000 feet (3,000 m) above sea level) and 700 hPa dew point.

CAPE and CIN

Convective available potential energy (CAPE), sometimes, simply, available potential energy (APE), is the amount of energy a parcel of air would have if lifted a certain distance vertically through the atmosphere. CAPE is effectively the positive buoyancy of an air parcel and is an indicator of atmospheric instability, which makes it valuable in predicting severe weather. CIN, convective inhibition, is effectively negative buoyancy, expressed B-; the opposite of convective available potential energy (CAPE), which is expressed as B+ or simply B. As with CAPE, CIN is usually expressed in J/kg but may also be expressed as m^2/s^2, as the values are equivalent. In fact, CIN is sometimes referred to as negative buoyant energy (NBE).

It is a form of fluid instability found in thermally stratified atmospheres in which a colder fluid overlies a warmer one. When an air mass is unstable, the element of the air mass that is displaced upwards is accelerated by the pressure differential between the displaced air and the ambient air at the (higher) altitude to which it was displaced. This usually creates vertically developed clouds from convection, due to the rising motion, which can eventually lead to thunderstorms. It could also be created in other phenomenon, such as a cold front. Even if the air is cooler on the surface, there is still warmer air in the mid-levels, that can rise into the upper-levels. However, if there is not enough water vapor present, there is no ability for condensation, thus storms, clouds, and rain will not form.

Bulk Richardson Number

The Bulk Richardson Number (BRN) is a dimensionless number relating vertical stability and vertical wind shear (generally, stability divided by shear). It represents the ratio of thermally-produced turbulence and turbulence generated by vertical shear. Practically, its value determines whether convection is free or forced. High values indicate unstable and/or weakly sheared environments; low values indicate weak instability and/or strong vertical shear. Generally, values in the range of around 10 to 45 suggest environmental conditions favorable for supercell development.

Showalter Index

The Showalter Index is a dimensionless number computed by taking the temperature at the 850 hPa level which is then taken dry adiabatically up to saturation, then up to the 500 hPa level, which is then subtracted by the observed 500 hPa level temperature. If the value is negative, then the lower portion of the atmosphere is unstable, with thunderstorms expected when the value is below -3. The application of Showalter Index is especially helpful when there is cool, shallow air mass below 850 hPa that conceals the potential convective lifting. However, the index will underestimate the potential convective lifting if there are cool layers that extend above 850 hPa and it does not consider diurnal radiative changes or moisture below 850 hPa.

Effects

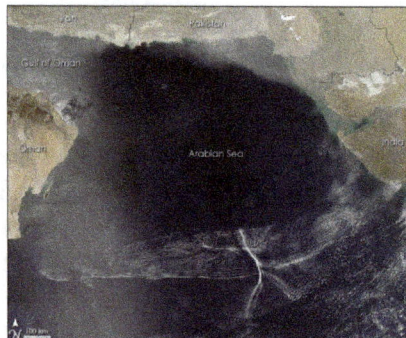

Image of an undular bore wave

Stable Atmosphere

Stable conditions, such as during a clear and calm night, will cause pollutants to become trapped near ground level. Drizzle occurs within a moist air mass when it is stable. Air within a stable layer is not turbulent. Conditions associated with a marine layer, a stable atmosphere common on the west side of continents near cold water currents, leads to overnight and morning fog. Undular bores can form when a low level boundary such as a cold front or outflow boundary approaches a layer of cold, stable air. The approaching boundary will create a disturbance in the atmosphere producing a wave-like motion, known as a gravity wave. Although the undular bore waves appear as bands of clouds across the sky, they are transverse waves, and are propelled by the transfer of energy from an on-coming storm and are shaped by gravity. The ripple like appearance of this wave is described as the disturbance in the water when a pebble is dropped into a pond or when a moving boat creates waves in the surrounding water. The object displaces the water or medium the wave is travelling through and the medium moves in an upward motion. However, because of gravity, the water or medium is pulled back down and the repetition of this cycle creates the transverse wave motion.

Unstable Atmosphere

Mirage over a hot road, with the appearance of "fake water" on its surface.

Within an unstable layer in the troposphere, the lifting of air parcels will occur, and continue for as long as the nearby atmosphere remains unstable. Once overturning through the depth of the troposphere occurs (with convection being capped by the relatively warmer, more stable layer of the stratosphere), deep convective currents lead to thunderstorm development when enough moisture is present. Over warm ocean waters and within a region of the troposphere with light vertical wind shear and significant low level spin (or vorticity), such thunderstorm activity can grow in coverage and develop into a tropical cyclone. Over hot surfaces during warm days, unstable dry air can lead to significant refraction of the light within the air layer, which causes inferior mirages.

When winds are light, dust devils can develop on dry days within a region of instability at ground level. Small-scale, tornado-like circulations can occur over or near any intense surface heat source, which would have significant instability in its vicinity. Those that occur near intense wildfires are called fire whirls, which can spread a fire beyond its previous bounds. A steam devil is a rotating updraft that involves steam or smoke. They can form from smoke issuing from a power plant smokestack. Hot springs and warm lakes are also suitable locations for a steam devil to form, when cold arctic air passes over the relatively warm water.

Lapse Rate

The lapse rate is the rate at which atmospheric temperature decreases with an increase in altitude. The terminology arises from the word *lapse* in the sense of a decrease or decline. While most often applied to Earth's troposphere, the concept can be extended to any gravitationally supported parcel of gas.

Definition

A formal definition from the *Glossary of Meteorology* is:

> The decrease of an atmospheric variable with height, the variable being temperature unless otherwise specified.

In general, a lapse rate is the negative of the rate of temperature change with altitude change, thus:

$$\gamma = -\frac{dT}{dz}$$

where γ is the lapse rate given in units of temperature divided by units of altitude, T = temperature, and z = altitude.

Convection and Adiabatic Expansion

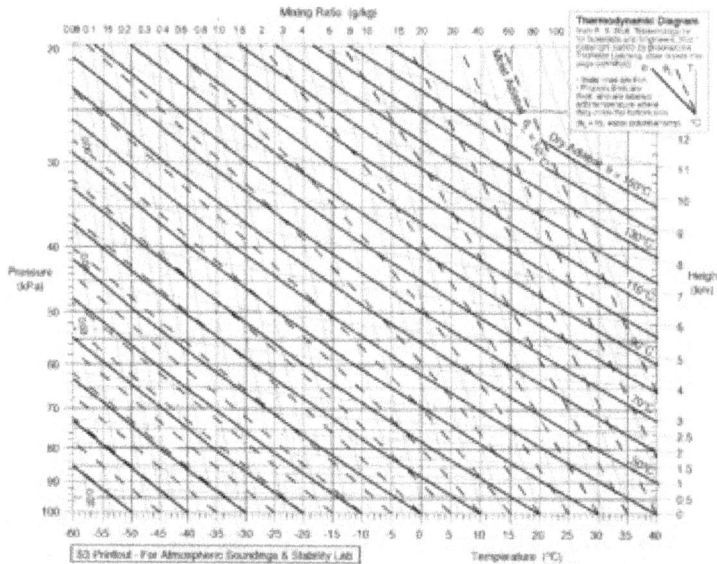

Emagram diagram showing variation of dry adiabats (bold lines) and moist adiabats (dash lines) according to pressure and temperature

The temperature profile of the atmosphere is a result of an interaction between radiation and convection. Sunlight hits the ground and heats it. The ground then heats the air at the surface. If radiation were the only way to transfer heat from the ground to space, the greenhouse effect of gases in the atmosphere would keep the ground at roughly 333 K (60°C; 140°F), and the temperature would decay exponentially with height.

However, when air is hot, it tends to expand, which lowers its density. Thus, hot air tends to rise and transfer heat upward. This is the process of convection. Convection comes to equilibrium when a parcel of air at a given altitude has the same density as the other air at the same elevation.

When a parcel of air expands, it pushes on the air around it, doing work (thermodynamics). Since the parcel does work but gains no heat, it loses internal energy so that its temperature decreases. The process of expanding and contracting without exchanging heat is an adiabatic process. The term *adiabatic* means that no heat transfer occurs into or out of the parcel. Air has low thermal conductivity, and the bodies of air involved are very large, so transfer of heat by conduction is negligibly small.

The adiabatic process for air has a characteristic temperature-pressure curve, so the process determines the lapse rate. When the air contains little water, this lapse rate is known as the dry adiabatic lapse rate: the rate of temperature decrease is 9.8°C/km (5.38°F per 1,000 ft) (3.0°C/1,000 ft). The reverse occurs for a sinking parcel of air.

Note that only the troposphere (up to approximately 12 kilometres (39,000 ft) of altitude) in the Earth's atmosphere undergoes convection: the stratosphere does not generally convect. However, some exceptionally energetic convection processes -- notably volcanic eruption columns and overshooting tops associated with severe supercell thunderstorms -- may *locally* and *temporarily* inject convection through the tropopause and into the stratosphere.

The mathematics of the adiabatic lapse rate can be derived from thermodynamics, which defines an adiabatic process via:

$$PdV = -VdP / \gamma$$

the first law of thermodynamics can be written as

$$mc_v dT - Vdp / \gamma = 0$$

Also since : $\alpha = V / m$ and : $\gamma = c_p / c_v$ we can show that:

$$c_p dT - \alpha dP = 0$$

where c_p is the specific heat at constant pressure and α is the specific volume.

Assuming an atmosphere in hydrostatic equilibrium:

$$dP = -\rho g dz$$

where g is the standard gravity and ρ is the density. Combining these two equations to eliminate the pressure, one arrives at the result for the dry adiabatic lapse rate (DALR),

$$\Gamma_d = -\frac{dT}{dz} = \frac{g}{c_p} = 9.8 \,^\circ C / km$$

Moist Adiabatic Lapse Rate

The presence of water within the atmosphere complicates the process of convection. Water vapor contains latent heat of vaporization. As air rises and cools, it eventually becomes saturated and cannot hold its quantity of water vapor. The water vapor condenses, forming clouds, and releasing heat. Before saturation, the rising air follows the dry adiabatic lapse rate. After saturation, the rising air follows the moist adiabatic lapse rate. The release of latent heat is an important source of energy in the development of thunderstorms.

While the dry adiabatic lapse rate is a constant 9.8°C/km (5.38°F per 1,000 ft, 3°C/1,000 ft), the moist adiabatic lapse rate varies strongly with temperature. A typical value is around 5°C/km, (9°F/km, 2.7°F/1,000 ft, 1.5°C/1,000 ft). The formula for the moist adiabatic lapse rate is given by:

$$\Gamma_w = g \frac{1 + \dfrac{H_v r}{R_{sd} T}}{c_{pd} + \dfrac{H_v^2 r}{R_{sw} T^2}} = g \frac{R_{sd} T^2 + H_v r T}{c_{pd} R_{sd} T^2 + H_v^2 r \epsilon}$$

where:	
	= Wet adiabatic lapse rate, K/m
g	= Earth's gravitational acceleration = 9.8076 m/s²
H_v	= Heat of vaporization of water, = 2501000 J/kg
R_{sd}	= Specific gas constant of dry air = 287 J kg⁻¹ K⁻¹
R_{sw}	= Specific gas constant of water vapour = 461.5 J kg⁻¹ K⁻¹
$\epsilon = \dfrac{R_{sd}}{R_{sw}}$	=The dimensionless ratio of the specific gas constant of dry air to the specific gas constant for water vapour = 0.622
e	= The water vapour pressure of the saturated air
p	= The pressure of the saturated air
$r = \epsilon e / (p - e)$	= The mixing ratio of the mass of water vapour to the mass of dry air
T	= Temperature of the saturated air, K
c_{pd}	= The specific heat of dry air at constant pressure, = 1003.5 J kg⁻¹ K⁻¹

Environmental Lapse Rate

The environmental lapse rate (ELR), is the rate of decrease of temperature with altitude in the stationary atmosphere at a given time and location. As an average, the International Civil Aviation Organization (ICAO) defines an international standard atmosphere (ISA) with a temperature lapse rate of 6.49 K/km (3.56°F or 1.98°C/1,000 ft) from sea level to 11 km (36,090 ft or 6.8 mi). From 11 km up to 20 km (65,620 ft or 12.4 mi), the constant temperature is −56.5°C (−69.7°F), which is the lowest assumed temperature in the ISA. The standard atmosphere contains no moisture. Unlike the idealized ISA, the temperature of the actual atmosphere does not always fall at a uniform rate with height. For example, there can be an inversion layer in which the temperature increases with altitude.

Effect on Weather

The varying environmental lapse rates throughout the Earth's atmosphere are of critical importance in meteorology, particularly within the troposphere. They are used to determine if the parcel of rising air will rise high enough for its water to condense to form clouds, and, having formed clouds, whether the air will continue to rise and form bigger shower clouds, and whether these clouds will get even bigger and form cumulonimbus clouds (thunder clouds).

As unsaturated air rises, its temperature drops at the dry adiabatic rate. The dew point also drops (as a result of decreasing air pressure) but much more slowly, typically about −2°C per 1,000 m. If unsaturated air rises far enough, eventually its temperature will reach its dew point, and condensation will begin to form. This altitude is known as the lifting condensation level (LCL) when mechanical lift is present and the convective condensation level (CCL) when mechanical lift is absent, in which case, the parcel must be heated from below to its convective temperature. The cloud base will be somewhere within the layer bounded by these parameters.

The difference between the dry adiabatic lapse rate and the rate at which the dew point drops is around 8°C per 1,000 m. Given a difference in temperature and dew point readings on the ground, one can easily find the LCL by multiplying the difference by 125 m/°C.

If the environmental lapse rate is less than the moist adiabatic lapse rate, the air is absolutely stable — rising air will cool faster than the surrounding air and lose buoyancy. This often happens in the early morning, when the air near the ground has cooled overnight. Cloud formation in stable air is unlikely.

If the environmental lapse rate is between the moist and dry adiabatic lapse rates, the air is conditionally unstable — an unsaturated parcel of air does not have sufficient buoyancy to rise to the LCL or CCL, and it is stable to weak vertical displacements in either direction. If the parcel is saturated it is unstable and will rise to the LCL or CCL, and either be halted due to an inversion layer of convective inhibition, or if lifting continues, deep, moist convection (DMC) may ensue, as a parcel rises to the level of free convection (LFC), after which it enters the free convective layer (FCL) and usually rises to the equilibrium level (EL).

If the environmental lapse rate is larger than the dry adiabatic lapse rate, it has a superadiabatic lapse rate, the air is absolutely unstable — a parcel of air will gain buoyancy as it rises both below and above the lifting condensation level or convective condensation level. This often happens in the afternoon mainly over land masses. In these conditions, the likelihood of cumulus clouds, showers or even thunderstorms is increased.

Meteorologists use radiosondes to measure the environmental lapse rate and compare it to the predicted adiabatic lapse rate to forecast the likelihood that air will rise. Charts of the environmental lapse rate are known as thermodynamic diagrams, examples of which include Skew-T log-P diagrams and tephigrams.

The difference in moist adiabatic lapse rate and the dry rate is the cause of foehn wind phenomenon (also known as "Chinook winds" in parts of North America). The phenomenon exists because warm moist air rises through orographic lifting up and over the top of a mountain range or large mountain. The temperature decreases with the dry adiabatic lapse rate, until it hits the dew point, where water vapor in the air begins to condense. Above that altitude, the adiabatic lapse rate decreases to the moist adiabatic lapse rate as the air continues to rise. Condensation is also commonly followed by precipitation on the top and windward sides of the mountain. As the air descends on the leeward side, it is warmed by adiabatic compression at the dry adiabatic lapse rate. Thus, the foehn wind at a certain altitude is warmer than the corresponding altitude on the windward side of the mountain range. In addition, because the air has lost much of its original water vapor content, the descending air creates an arid region on the leeward side of the mountain.

Bulk Richardson Number

The Bulk Richardson Number (BRN) is an approximation of the Gradient Richardson number. The BRN is a dimensionless ratio in meteorology related to the consumption of turbulence divided by the shear production (the generation of turbulence kinetic energy caused by wind shear) of turbulence. It is used to show dynamic stability and the formation of turbulence.

The BRN is used frequently in meteorology due to widely available rawinsonde (frequently called radiosonde) data and numerical weather forecasts that supply wind and temperature measurements at discrete points in space.

Formula

Below is the formula for the BRN. Where g is gravitational acceleration, T_v is absolute virtual temperature, $\Delta\theta_v$ is the virtual potential temperature difference across a layer of thickness Δz (vertical depth), and ΔU and ΔV are the changes in horizontal wind components across that same layer.

$$R_B = \frac{(g/T_v)\Delta\theta_v\Delta z}{(\Delta U)^2 + (\Delta V)^2}$$

Critical Values and Interpretation

High values indicate unstable and/or weakly-sheared environments; low values indicate weak instability and/or strong vertical shear. Generally, values in the range of around 50 to 100 suggest environmental conditions favorable for supercell development.

In the limit of layer thickness becoming small, the Bulk Richardson number approaches the Gradient Richardson number, for which a critical Richardson number is roughly $Ri_c = 0.25$. Numbers less than this critical value are dynamically unstable and likely to become or remain turbulent.

The critical value of 0.25 applies only for local gradients, not for finite differences across thick layers. The thicker the layer is the more likely we are to average out large gradients that occur within small sub-regions of the layer of interest. This results in uncertainty of our prediction of the occurrence of turbulence, and now one must use an artificially large value of the critical Richardson number to give reasonable results using our smoothed gradients. This means that the thinner the layer, the closer the value to the theory.

Convective Available Potential Energy

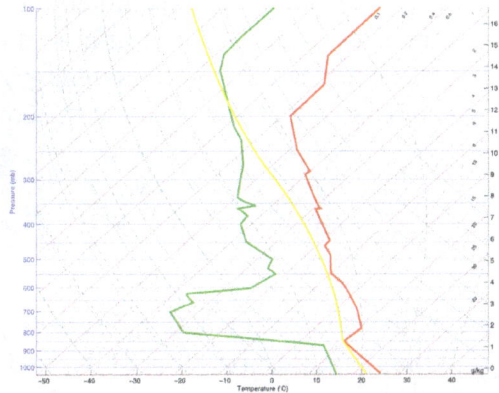

A skew-T plot showing a morning sounding with a large hydrolapse followed by an afternoon sounding showing the cooling (red curve moving to the left) which occurred in the mid-levels resulting in an unstable atmosphere as surface parcels have now become negatively buoyant. The red line is temperature, the green line is the dew point, and the yellow line is the air parcel lifted.

In meteorology, convective available potential energy (CAPE), sometimes, simply, available potential energy (APE), is the amount of energy a parcel of air would have if lifted a certain distance vertically through the atmosphere. CAPE is effectively the positive buoyancy of an air parcel and is an indicator of atmospheric instability, which makes it very valuable in predicting severe weather. It is a form of fluid instability found in thermally stratified atmospheres in which a colder fluid overlies a warmer one. As explained below, when an air mass is unstable, the element of the air mass that is displaced upwards is accelerated by the pressure differential between the displaced air and the ambient air at the (higher) altitude to which it was displaced. This usually creates vertically developed clouds from convection, due to the rising motion, which can eventually lead to thunderstorms. It could also be created by other phenomena, such as a cold front. Even if the air is cooler on the surface, there is still warmer air in the mid-levels, that can rise into the upper-levels. However, if there is not enough water vapor present, there is no ability for condensation, thus storms, clouds, and rain will not form.

Mechanics

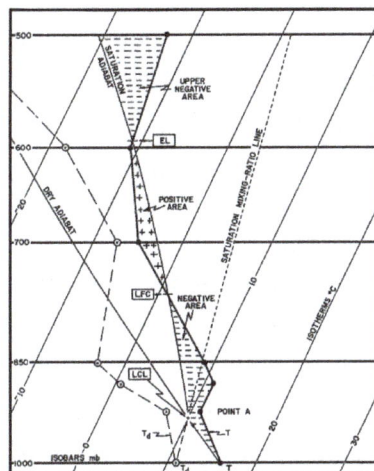

A Skew-T diagram with important features labeled

CAPE exists within the conditionally unstable layer of the troposphere, the free convective layer (FCL), where an ascending air parcel is warmer than the ambient air. CAPE is measured in joules per kilogram of air (J/kg). Any value greater than 0 J/kg indicates instability and an increasing possibility of thunderstorms and hail. Generic CAPE is calculated by integrating vertically the local buoyancy of a parcel from the level of free convection (LFC) to the equilibrium level (EL):

$$\text{CAPE} = \int_{z_f}^{z_n} g \left(\frac{T_{v,\text{parcel}} - T_{v,\text{env}}}{T_{v,\text{env}}} \right) dz$$

Where z_f is the height of the level of free convection and z_n is the height of the equilibrium level(-neutral buoyancy), where $T_{v,\text{parcel}}$ is the virtual temperature of the specific parcel, where $T_{v,\text{env}}$ is the virtual temperature of the environment, and where g is the acceleration due to gravity. CAPE for a given region is most often calculated from a thermodynamic or sounding diagram (e.g., a Skew-T log-P diagram) using air temperature and dew point data usually measured by a weather balloon.

CAPE is effectively positive buoyancy, expressed $B+$ or simply B; the opposite of convective inhibition (CIN), which is expressed as $B-$, and can be thought of as "negative CAPE". As with CIN, CAPE is usually expressed in J/kg but may also be expressed as m^2/s^2, as the values are equivalent. In fact, CAPE is sometimes referred to as *positive buoyant energy* (*PBE*). This type of CAPE is the maximum energy available to an ascending parcel and to moist convection. When a layer of CIN is present, the layer must be eroded by surface heating or mechanical lifting, so that convective boundary layer parcels may reach their level of free convection (LFC).

On a sounding diagram, CAPE is the *positive area* above the LFC, the area between the parcel's virtual temperature line and the environmental virtual temperature line where the ascending parcel is warmer than the environment. Neglecting the virtual temperature correction may result in substantial relative errors in the calculated value of CAPE for small CAPE values. CAPE may also exist below the LFC, but if a layer of CIN (subsidence) is present, it is unavailable to deep, moist convection until CIN is exhausted. When there is mechanical lift to saturation, cloud base begins at the lifted condensation level (LCL); absent forcing, cloud base begins at the convective condensation level (CCL) where heating from below causes spontaneous buoyant lifting to the point of condensation when the convective temperature is reached. When CIN is absent or is overcome, saturated parcels at the LCL or CCL, which had been small cumulus clouds, will rise to the LFC, and then spontaneously rise until hitting the stable layer of the equilibrium level. The result is deep, moist convection (DMC), or simply, a thunderstorm.

When a parcel is unstable, it will continue to move vertically, in either direction, dependent on whether it receives upward or downward forcing, until it reaches a stable layer (though momentum, gravity, and other forcing may cause the parcel to continue). There are multiple types of CAPE, *downdraft CAPE* (*DCAPE*), estimates the potential strength of rain and evaporatively cooled downdrafts. Other types of CAPE may depend on the depth being considered. Other examples are *surface based CAPE* (*SBCAPE*), *mixed layer* or *mean layer CAPE* (*MLCAPE*), *most unstable* or *maximum usable CAPE* (*MUCAPE*), and *normalized CAPE* (*NCAPE*).

Fluid elements displaced upwards or downwards in such an atmosphere expand or compress adiabatically in order to remain in pressure equilibrium with their surroundings, and in this manner become less or more dense.

If the adiabatic decrease or increase in density is *less* than the decrease or increase in the density of the ambient (not moved) medium, then the displaced fluid element will be subject to downwards or upwards pressure, which will function to restore it to its original position. Hence, there will be a counteracting force to the initial displacement. Such a condition is referred to as *convective stability*.

On the other hand, if adiabatic decrease or increase in density is *greater* than in the ambient fluid, the upwards or downwards displacement will be met with an additional force in the same direction exerted by the ambient fluid. In these circumstances, small deviations from the initial state will become amplified. This condition is referred to as *convective instability*.

Convective instability is also termed *static instability*, because the instability does not depend on the existing motion of the air; this contrasts with dynamic instability where instability is dependent on the motion of air and its associated effects such as dynamic lifting.

Significance to Thunderstorms

Thunderstorms form when air parcels are lifted vertically. Deep, moist convection requires a parcel to be lifted to the LFC where it then rises spontaneously until reaching a layer of non-positive buoyancy. The atmosphere is warm at the surface and lower levels of the troposphere where there is mixing (the planetary boundary layer (PBL)), but becomes substantially cooler with height. The temperature profile of the atmosphere, the change in temperature, the degree that it cools with height, is the lapse rate. When the rising air parcel cools more slowly than the surrounding atmosphere, it remains warmer and less dense. The parcel continues to rise freely (convectively; without mechanical lift) through the atmosphere until it reaches an area of air less dense (warmer) than itself.

The amount, and shape, of the positive-buoyancy area modulates the speed of updrafts, thus extreme CAPE can result in explosive thunderstorm development; such rapid development usually occurs when CAPE stored by a capping inversion is released when the "lid" is broken by heating or mechanical lift. The amount of CAPE also modulates how low-level vorticity is entrained and then stretched in the updraft, with importance to tornadogenesis. The most important CAPE for tornadoes is within the lowest 1 to 3 km (0.6 to 1.9 mi) of the atmosphere, whilst deep layer CAPE and the width of CAPE at mid-levels is important for supercells. Tornado outbreaks tend to occur within high CAPE environments. Large CAPE is required for the production of very large hail, owing to updraft strength, although a rotating updraft may be stronger with less CAPE. Large CAPE also promotes lightning activity.

Two notable days for severe weather exhibited CAPE values over 5 kJ/kg. Two hours before the 1999 Oklahoma tornado outbreak occurred on May 3, 1999, the CAPE value sounding at Oklahoma City was at 5.89 kJ/kg. A few hours later, an F5 tornado ripped through the southern suburbs of the city. Also on May 4, 2007 CAPE values of 5.5 kJ/kg were reached and an EF5 tornado tore through Greensburg, Kansas. On these days, it was apparent that conditions were ripe for tornadoes and CAPE wasn't a crucial factor. However, extreme CAPE, by modulating the updraft (and downdraft), can allow for exceptional events, such as the deadly F5 tornadoes that hit Plainfield, Illinois on August 28, 1990 and Jarrell, Texas on May 27, 1997 on days which weren't readily apparent as conducive to large tornadoes. CAPE was estimated to exceed 8 kJ/kg in the environment of the Plainfield storm and was around 7 kJ/kg for the Jarrell storm.

Severe weather and tornadoes can develop in an area of low CAPE values. The surprise severe weather event that occurred in Illinois and Indiana on April 20, 2004 is a good example. Importantly in that case, was that although overall CAPE was weak, there was strong CAPE in the lowest levels of the troposphere which enabled an outbreak of minisupercells producing large, long-track, intense tornadoes.

Example from Meteorology

A good example of convective instability can be found in our own atmosphere. If dry mid-level air is drawn over very warm, moist air in the lower troposphere, a hydrolapse (an area of rapidly decreasing dew point temperatures with height) results in the region where the moist boundary layer and mid-level air meet. As daytime heating increases mixing within the moist boundary layer, some of the moist air will begin to interact with the dry mid-level air above it. Owing to thermodynamic processes, as the dry mid-level air is slowly saturated its temperature begins to drop, increasing the adiabatic lapse rate. Under certain conditions, the lapse rate can increase significantly in a short amount of time, resulting in convection. High convective instability can lead to severe thunderstorms and tornadoes as moist air which is trapped in the boundary layer eventually becomes highly negatively buoyant relative to the adiabatic lapse rate and escapes as a rapidly rising bubble of humid air triggering the development of a cumulus or cumulonimbus cloud.

Atmospheric Sounding

An atmospheric sounding is a measurement of vertical distribution of physical properties of the atmospheric column such as pressure, temperature, wind speed and wind direction (thus deriving wind shear), liquid water content, ozone concentration, pollution, and other properties. Such measurements are performed in a variety of ways including remote sensing and in situ observations.

The most common in situ sounding is a radiosonde, which usually is a weather balloon, but can also be a rocketsonde.

Remote sensing soundings generally use passive infrared and microwave radiometers:

- airborne instruments
- surface stations
- Earth-observing satellite instruments such as AIRS and AMSU
- observation of atmospheres on different planets, such as the Mars climate sounder on the Mars Reconnaissance Orbiter

Direct Methods

Sensors that measure atmospheric constituents directly, such as thermometers, barometers, and humidity sensors, can be sent aloft on balloons, rockets or dropsondes. They can also be carried on the outer hulls of ships and aircraft or even mounted on towers. In this case, all that is needed to capture the measurements are storage devices and/or transponders.

Indirect Methods

The more challenging case involves sensors, primarily satellite-mounted, such as radiometers, optical sensors, radar, lidar and ceilometer as well as sodar since these cannot measure the quantity of interest, such as temperature, pressure, humidity etc., directly. By understanding emission and absorption processes, we can figure out what the instrument is looking at between the layers of atmosphere. While this type of instrument can also be operated from ground stations or vehicles—optical methods can also be used inside in situ instruments—satellite instruments are particularly important because of their extensive, regular coverage. The AMSU instruments on three NOAA and two EUMETSAT satellites, for instance, can sample the entire globe at better than one degree resolution in less than a day.

We can distinguish between two broad classes of sensor: *active*, such as radar, that have their own source, and *passive* that only detect what is already there. There can be a variety of sources for a passive instrument, including scattered radiation, light emitted directly from the sun, moon or stars—both more appropriate in the visual or ultra-violet range—as well light emitted from warm objects, which is more appropriate in the microwave and infrared.

Viewing Geometry

A limb sounder looks at the edge of the atmosphere where it is visible above the Earth. It does this in one of two ways: either it tracks the sun, moon, a star, or another transmitting satellite through the limb as the source gets occultated behind the Earth, or it looks towards empty space, collecting radiation that is scattered from one of these sources. In contrast, a nadir-looking atmospheric sounder looks down through the atmosphere at the surface. The SCIAMACHY instrument operates in all three of these modes.

Atmospheric Inverse Problem

Statement of the Problem

The following applies mainly to passive sensors, but has some applicability to active sensors.

Typically, there is a vector of values of the quantity to be retrieved, $\vec{x}$, called the state vector and a vector of measurements, $\vec{y}$. The state vector could be temperatures, ozone number densities, humidities etc. The measurement vector is typically counts, radiances or brightness temperatures from a radiometer or similar detector but could include any other quantity germain to the problem. The forward model maps the state vector to the measurement vector:

$$\vec{y} = \vec{f}(\vec{x})$$

Usually the mapping, $\vec{f}$, is known from physical first principles, but this may not always be the case. Instead, it may only be known empirically, by matching actual measurements with actual states. Satellite and many other remote sensing instruments do not measure the relevant physical properties, that is the state, but rather the amount of radiation emitted in a particular direction, at a particular frequency. It is usually easy to go from the state space to the measurement space—for instance with Beer's law or radiative transfer—but not the other way around, therefore we need some method of inverting $\vec{f}$ or of finding the inverse model, $\vec{f}^{-1}$.

Methods of Solution

If the problem is linear we can use some type of matrix inverse method—often the problem is ill-posed or unstable so we will need to regularize it: good, simple methods include the normal equation or singular value decomposition. If the problem is weakly nonlinear, an iterative method such Newton-Raphson may be appropriate.

Sometimes the physics is too complicated to model accurately or the forward model too slow to be used effectively in the inverse method. In this case, statistical or machine learning methods such as linear regression, neural networks, statistical classification, kernel estimation, etc. can be used to form an inverse model based on a collection of ordered pairs of samples mapping the state space to the measurement space, that is, $\{\vec{x}:\vec{y}\}$. These can be generated either from models—e.g. state vectors from dynamical models and measurement vectors from radiative transfer or similar forward models—or from direct, empirical measurement. Other times when a statistical method might be more appropriate include highly nonlinear problems.

List of Methods

- Differential absorption spectroscopy

- Isoline retrieval

- Optimal estimation

Atmospheric Temperature Range

Atmospheric temperature range is the numerical difference between the minimum and maximum values of temperature observed in a given location during a period of time (e.g., in a given day, month, year, century) or the average (average of all temperature ranges in a period of time). The variation in temperature that occurs from the highs of the day to the cool of nights is called diurnal temperature variation.

Parameters Affecting the Range

The size of ground-level atmospheric temperature ranges depends on several factors, such as:

- The average temperature

- The average humidity

- The regime of winds (intensity, duration, variation, temperature, etc.)

- The proximity to large bodies of water, such as the sea

The figure below shows an example of monthly temperatures recorded at one of such locations, the city of Campinas, state of São Paulo, Brazil, which lies approximately 60 km north of the Capricorn line (latitude of 22 degrees). Average yearly temperature is 22.4 degrees Celsius, ranging from an average minimum of 12.2 degrees to a maximum of 29.9 degrees. The average temperature range is 11.4 degrees. Variability along the year is small (standard deviation of 2.31 for the maximum monthly average and 4.11 for the minimum). It is easy to see in the graph another typical phenomenon of temperature ranges, which is its increase during winter (lower average air temperature).

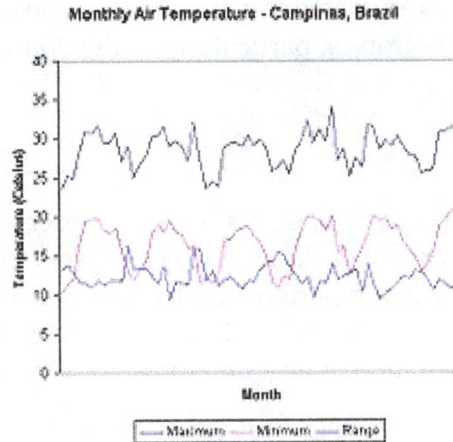

Average maximum, minimum and range of monthly air temperatures recorded in Campinas, Brazil, between January 2001 and July 2006.

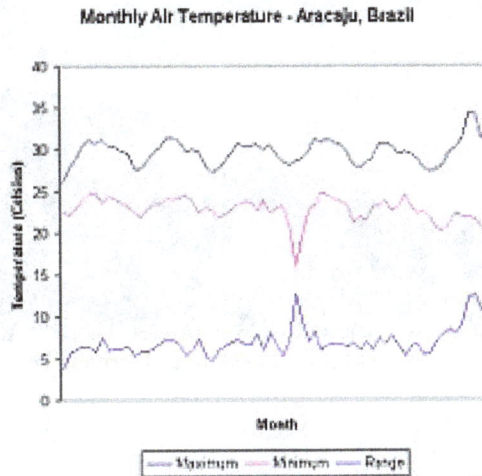

Average maximum, minimum and range of monthly air temperatures recorded in Aracaju, state of Sergipe, Brazil, between January 2001 and July 2006.

In Campinas, for example, the daily temperature range in July (the coolest month of the year) may vary between typically 10 and 24 degrees Celsius (range of 14), while in January, it may range between 20 and 30 degrees Celsius (range of 10).

The effect of latitude, tropical climate, constant gentle wind and sea-side locations show smaller average temperature ranges, smaller variations of temperature, and a higher average temperature (second graph, taken for the same period as Campinas, at Aracaju, capital of the state of Sergipe, also in Brazil, at a latitude of 10 degrees, nearer to the Equator). Average maximum yearly temperature is 28.7 degrees Celsius and average minimum is 21.9. The average temperature range is 5.7 degrees only. Temperature variation along the year in Aracaju is very damped (standard deviation of 1.93 for the maximum temperature and 2.72 for the minimum temperature).

Uses

A location which combines an average temperature of 19 degrees Celsius, 60% average humidity and a temperature range of about 10 degrees Celsius around the average temperature (yearly temperature variation) is considered ideal in terms of comfort for the human species. Most of the places with these characteristics are located in the transition between temperate and tropical climates, approximately around the tropics, particularly in the Southern hemisphere (the tropic of Capricorn).

Lifted Minimum Temperature

The minimum temperature on calm, clear nights has been observed to occur not on the ground, but rather a few tens of centimeters above the ground. The lowest temperature layer is called *Ramdas layer* after L. A. Ramdas, who first reported this phenomenon in 1932 based on observations at different screen heights at six meteorological centers across India. The phenomenon is attributed to the interaction of thermal radiation effects on atmospheric aerosols and convection transfer close to the ground.

Inversion (Meteorology)

Smoke rising in Lochcarron, Scotland, is stopped by an overlying layer of warmer air (2006).

A temperature inversion in Budapest, Hungary viewing Margaret Island - 2013.

In meteorology, an inversion is a deviation from the normal change of an atmospheric property with altitude.

It almost always refers to a "temperature inversion", i.e. an increase in temperature with height, or to the layer ("inversion layer") within which such an increase occurs.

An inversion can lead to pollution such as smog being trapped close to the ground, with possible adverse effects on health. An inversion can also suppress convection by acting as a "cap". If this cap is broken for any of several reasons, convection of any moisture present can then erupt into violent thunderstorms. Temperature inversion can notoriously result in freezing rain in cold climates.

Normal Atmospheric Conditions

Usually, within the lower atmosphere (the troposphere) the air near the surface of the Earth is warmer than the air above it, largely because the atmosphere is heated from below as solar radiation warms the Earth's surface, which in turn then warms the layer of the atmosphere directly above it, e.g., by thermals (convective heat transfer).

Causes

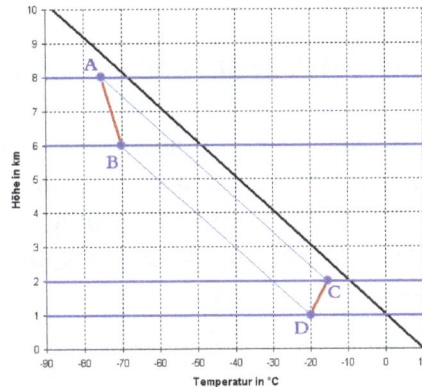

Height (y-axis) versus temperature (x-axis) under normal atmospheric conditions (black line). When the layer from 6–8 kilometres (4–5 miles) (designated A-B) descends dry adiabatically , the result is the inversion seen near the ground at 1–2 kilometres (1–1 mile) (C-D).

Klagenfurter Becken in December 2015: on mount Goritschnigkogel there is a distinct inverse hoarfrost margin.

Given the right conditions, the normal vertical temperature gradient is inverted such that the air is colder near the surface of the Earth. This can occur when, for example, a warmer, less-dense air mass moves over a cooler, denser air mass. This type of inversion occurs in the vicinity of warm fronts, and also in areas of oceanic upwelling such as along the California coast in the United States. With sufficient humidity in the cooler layer, fog is typically present below the inversion cap. An inversion is also produced whenever radiation from the surface of the earth exceeds the amount of radiation received from the sun, which commonly occurs at night, or during the winter when the angle of the sun is very low in the sky. This effect is virtually confined to land regions as the ocean retains heat far longer. In the polar regions during winter, inversions are nearly always present over land.

A warmer air mass moving over a cooler one can "shut off" any convection which may be present in the cooler air mass. This is known as a capping inversion. However, if this cap is broken, either by extreme convection overcoming the cap, or by the lifting effect of a front or a mountain range, the sudden release of bottled-up convective energy – like the bursting of a balloon – can result in severe thunderstorms. Such capping inversions typically precede the development of tornadoes in the Midwestern United States. In this instance, the "cooler" layer is actually quite warm, but is still denser and usually cooler than the lower part of the inversion layer capping it.

Subsidence Inversion

An inversion can develop aloft as a result of air gradually sinking over a wide area and being warmed by adiabatic compression, usually associated with subtropical high-pressure areas. A stable marine layer may then develop over the ocean as a result. As this layer moves over progressively warmer waters, however, turbulence within the marine layer can gradually lift the inversion layer to higher altitudes, and eventually even pierce it, producing thunderstorms, and under the right circumstances, tropical cyclones. The accumulated smog and dust under the inversion quickly taints the sky reddish, easily seen on sunny days.

Consequences

A Fata Morgana (or mirage) of a ship is due to an inversion (2008).

Winter smoke in Shanghai, China, with a clear border-layer for the vertical air-spread (1993).

A temperature inversion in Bratislava, Slovakia, viewing the top of Nový Most (2005).

Temperature inversion stops atmospheric convection (which is normally present) from happening in the affected area and can lead to the air becoming stiller and murky from the collection of dust and pollutants that are no longer able to be lifted from the surface. This can become a problem in cities where many pollutants exist. Inversion effects occur frequently in big cities such as:

- Beijing, China
- Chengdu, China
- Los Angeles, United States
- Mexico City, Mexico
- Milan, Italy
- Monterrey, Mexico
- Mumbai, India
- San Francisco, United States
- Santiago, Chile
- São Paulo, Brazil
- Tehran, Iran

but also in smaller cities such as:

- Belgrade, Serbia
- Bergen, Norway
- Berkeley, California, United States
- Boise, Idaho, United States
- Bratislava, Slovakia
- Brisbane, Australia

- Budapest, Hungary

- Chiang Mai, Thailand

- Christchurch, New Zealand

- Córdoba, Argentina

- Fairbanks, Alaska, United States

- Launceston, Tasmania, Australia

- Ljubljana, Slovenia

- Logan, Utah, United States

- Oslo, Norway

- Prague, Czech Republic

- Provo, Utah, United States

- Salt Lake City, United States

- Sarajevo, Bosnia and Herzegovina

- Skopje, Republic of Macedonia

- Sofia, Bulgaria

- Tabriz, Iran

- Vancouver, Canada

- Venice, Italy

- Vienna, Austria

These cities are closely surrounded by hills and mountains, or on plains which are surrounded by mountain chains, which makes an inversion trap the air in the city. During a severe inversion, trapped air pollutants form a brownish haze that can cause respiratory problems. The Great Smog of 1952 in London, England, is one of the most serious examples of such an inversion. It was blamed for an estimated 11,000 to 12,000 deaths.

Sometimes the inversion layer is at a high enough altitude that cumulus clouds can condense but can only spread out under the inversion layer. This decreases the amount of sunlight reaching the ground and prevents new thermals from forming. As the clouds disperse, sunny weather replaces cloudiness in a cycle that can occur more than once a day.

As the temperature of air increases, the index of refraction of air decreases, a side effect of hotter air being less dense. Normally this results in distant objects being shortened vertically, an effect that is easy to see at sunset where the sun is visible as an oval. In an inversion, the normal pattern is reversed, and distant objects are instead stretched out or appear to be above the horizon, leading to the phenomenon known as a Fata Morgana or mirage.

Electromagnetic Radiation (Radio and Television)

Very high frequency radio waves can be refracted by inversions, making it possible to hear FM radio or watch VHF low-band television broadcasts from long distances on foggy nights. The signal, which would normally be refracted up and away from the ground-based antenna, is instead refracted down towards the earth by the temperature-inversion boundary layer. This phenomenon is called tropospheric ducting. Along coast lines during Autumn and Spring, due to multiple stations being simultaneously present because of reduced propagation losses, many FM radio stations are plagued by severe signal degradation causing them to sound scrambled.

Inversions can magnify the so-called "green flash": a phenomenon occurring at sunrise or sunset, usually visible for a few seconds, in which the sun's green light is isolated due to dispersion – the shorter wavelength is refracted most, so it is the first or last light from the upper rim of the solar disc to be seen.

Sound

When an inversion layer is present, if a sound or explosion occurs at ground level, the sound wave is reflected from the warmer upper layer and returns towards the ground. The sound, therefore, travels much farther than normal. This is noticeable in areas around airports, where the sound of aircraft taking off and landing often can be heard at greater distances around dawn than at other times of day, and inversion thunder which is significantly louder and travels further than when it is produced by lightning strikes under normal conditions.

Shock Waves

The shock wave from an explosion can be reflected by an inversion layer in much the same way as it bounces off the ground in an air-burst and can cause additional damage as a result. This phenomenon killed three people in the Soviet RDS-37 nuclear test when a building collapsed.

Thermodynamic Temperature

Thermodynamic temperature is the absolute measure of temperature and is one of the principal parameters of thermodynamics.

Thermodynamic temperature is defined by the third law of thermodynamics in which the theoretically lowest temperature is the null or zero point. At this point, absolute zero, the particle constituents of matter have minimal motion and can become no colder. In the quantum-mechanical description, matter at absolute zero is in its ground state, which is its state of lowest energy. Thermodynamic temperature is often also called absolute temperature, for two reasons: one, proposed by Kelvin, that it does not depend on the properties of a particular material; two that it refers to an absolute zero according to the properties of the ideal gas.

The International System of Units specifies a particular scale for thermodynamic temperature. It uses the kelvin scale for measurement and selects the triple point of water at 273.16 K as the fundamental fixing point. Other scales have been in use historically. The Rankine scale, using the degree Fahrenheit as its unit interval, is still in use as part of the English Engineering Units in the United

States in some engineering fields. ITS-90 gives a practical means of estimating the thermodynamic temperature to a very high degree of accuracy.

Roughly, the temperature of a body at rest is a measure of the mean of the energy of the translational, vibrational and rotational motions of matter's particle constituents, such as molecules, atoms, and subatomic particles. The full variety of these kinetic motions, along with potential energies of particles, and also occasionally certain other types of particle energy in equilibrium with these, make up the total internal energy of a substance. Internal energy is loosely called the heat energy or thermal energy in conditions when no work is done upon the substance by its surroundings, or by the substance upon the surroundings. Internal energy may be stored in a number of ways within a substance, each way constituting a "degree of freedom". At equilibrium, each degree of freedom will have on average the same energy: $k_B T / 2$ where k_B is the Boltzmann constant, unless that degree of freedom is in the quantum regime. The internal degrees of freedom (rotation, vibration, etc.) may be in the quantum regime at room temperature, but the translational degrees of freedom will be in the classical regime except at extremely low temperatures (fractions of kelvins) and it may be said that, for most situations, the thermodynamic temperature is specified by the average translational kinetic energy of the particles.

Overview

Temperature is a measure of the random submicroscopic motions and vibrations of the particle constituents of matter. These motions comprise the internal energy of a substance. More specifically, the thermodynamic temperature of any bulk quantity of matter is the measure of the average kinetic energy per classical (i.e., non-quantum) degree of freedom of its constituent particles. "Translational motions" are almost always in the classical regime. Translational motions are ordinary, whole-body movements in three-dimensional space in which particles move about and exchange energy in collisions. Thermodynamic temperature's null point, absolute zero, is the temperature at which the particle constituents of matter are as close as possible to complete rest; that is, they have minimal motion, retaining only quantum mechanical motion. Zero kinetic energy remains in a substance at absolute zero.

Throughout the scientific world where measurements are made in SI units, thermodynamic temperature is measured in kelvins (symbol: K). Many engineering fields in the U.S. however, measure thermodynamic temperature using the Rankine scale.

By international agreement, the unit *kelvin* and its scale are defined by two points: absolute zero, and the triple point of Vienna Standard Mean Ocean Water (water with a specified blend of hydrogen and oxygen isotopes). Absolute zero, the lowest possible temperature, is defined as being precisely 0 K *and* −273.15°C. The triple point of water is defined as being precisely 273.16 K *and* 0.01°C. This definition does three things:

1. It fixes the magnitude of the kelvin unit as being precisely 1 part in 273.16 parts the difference between absolute zero and the triple point of water;

2. It establishes that one kelvin has precisely the same magnitude as a one-degree increment on the Celsius scale; and

3. It establishes the difference between the two scales' null points as being precisely 273.15 kelvins (0 K = −273.15°C and 273.16 K = 0.01°C).

Temperatures expressed in kelvins are converted to degrees Rankine simply by multiplying by 1.8 as follows: $T_{°R} = 1.8T_K$, where T_K and $T_{°R}$ are temperatures in kelvins and degrees Rankine respectively. Temperatures expressed in degrees Rankine are converted to kelvins by *dividing* by 1.8 as follows: $T_K = {}^{T_{°R}}/_{1.8}$.

Practical Realization

Although the kelvin and Celsius scales are defined using absolute zero (0 K) and the triple point of water (273.16 K and 0.01°C), it is impractical to use this definition at temperatures that are very different from the triple point of water. ITS-90 is then designed to represent the thermodynamic temperature as closely as possible throughout its range. Many different thermometer designs are required to cover the entire range. These include helium vapor pressure thermometers, helium gas thermometers, standard platinum resistance thermometers (known as SPRTs, PRTs or Platinum RTDs) and monochromatic radiation thermometers.

For some types of thermometer the relationship between the property observed (e.g., length of a mercury column) and temperature, is close to linear, so for most purposes a linear scale is sufficient, without point-by-point calibration. For others a calibration curve or equation is required. The mercury thermometer, invented before the thermodynamic temperature was understood, originally *defined* the temperature scale; its linearity made readings correlate well with true temperature, i.e. the "mercury" temperature scale was a close fit to the true scale.

The Relationship of Temperature, Motions, Conduction, and Thermal Energy

The nature of Kinetic Energy, Translational Motion, and Temperature

The thermodynamic temperature is a measure of the average energy of the translational, vibrational, and rotational motions of matter's particle constituents (molecules, atoms, and subatomic particles). The full variety of these kinetic motions, along with potential energies of particles, and also occasionally certain other types of particle energy in equilibrium with these, contribute the total internal energy (loosely, the thermal energy) of a substance. Thus, internal energy may be stored in a number of ways (degrees of freedom) within a substance. When the degrees of freedom are in the classical regime ("unfrozen") the temperature is very simply related to the average energy of those degrees of freedom at equilibrium. The three translational degrees of freedom are unfrozen except for the very lowest temperatures, and their kinetic energy is simply related to the thermodynamic temperature over the widest range. The heat capacity, which relates heat input and temperature change, is discussed below.

The relationship of kinetic energy, mass, and velocity is given by the formula $E_k = \frac{1}{2}mv^2$. Accordingly, particles with one unit of mass moving at one unit of velocity have precisely the same kinetic energy, and precisely the same temperature, as those with four times the mass but half the velocity.

Except in the quantum regime at extremely low temperatures, the thermodynamic temperature of any *bulk quantity* of a substance (a statistically significant quantity of particles) is directly proportional to the mean average kinetic energy of a specific kind of particle motion known as *translational motion*. These simple movements in the three x, y, and z–axis dimensions of space means the particles move in the three spatial *degrees of freedom*. The temperature derived from this translational kinetic energy is sometimes referred to as *kinetic temperature* and is equal to the thermodynamic temperature over a very wide range of temperatures. Since there are three translational degrees of freedom (e.g., motion along the x, y, and z axes), the translational kinetic energy is related to the kinetic temperature by:

$$\bar{E} = \frac{3}{2} k_B T_k$$

where:

- $\bar{E}$ is the mean kinetic energy in joules (J) and is pronounced "E bar"

- $k_B = 1.3806504(24) \times 10^{-23}$ J/K is the Boltzmann constant and is pronounced "Kay sub bee"

- T_k is the kinetic temperature in kelvins (K) and is pronounced "Tee sub kay"

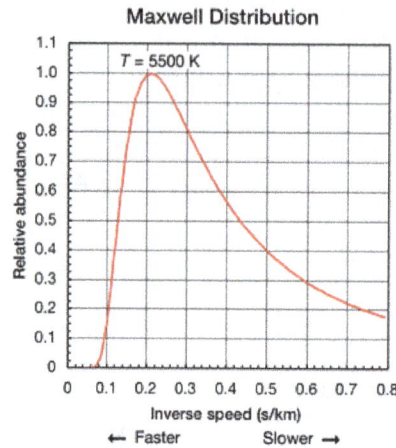

The translational motions of helium atoms occur across a range of speeds. Compare the shape of this curve to that of a Planck curve.

While the Boltzmann constant is useful for finding the mean kinetic energy of a particle, it's important to note that even when a substance is isolated and in thermodynamic equilibrium (all parts are at a uniform temperature and no heat is going into or out of it), the translational motions of individual atoms and molecules occur across a wide range of speeds. At any one instant, the proportion of particles moving at a given speed within this range is determined by probability as described by the Maxwell–Boltzmann distribution. The graph shown here in the *figure above* shows the speed distribution of 5500 K helium atoms. They have a *most probable* speed of 4.780 km/s. However, a certain proportion of atoms at any given instant are moving faster while others are moving relatively slowly; some are momentarily at a virtual standstill (off the x–axis to the right). This graph uses *inverse speed* for its x–axis so the shape of the curve can easily be compared to the curves in *figure* below. In both graphs, zero on the x–axis represents infinite temperature. Additionally, the x and y–axis on both graphs are scaled proportionally.

The High Speeds of Translational Motion

Although very specialized laboratory equipment is required to directly detect translational motions, the resultant collisions by atoms or molecules with small particles suspended in a fluid produces Brownian motion that can be seen with an ordinary microscope. The translational motions of elementary particles are *very* fast and temperatures close to absolute zero are required to directly observe them. For instance, when scientists at the NIST achieved a record-setting cold temperature of 700 nK (billionths of a kelvin) in 1994, they used optical lattice laser equipment to adiabatically cool caesium atoms. They then turned off the entrapment lasers and directly measured atom velocities of 7 mm per second in order to calculate their temperature. Formulas for calculating the velocity and speed of translational motion are given in the following footnote.

The Internal Motions of Molecules and Specific Heat

There are other forms of internal energy besides the kinetic energy of translational motion. As can be seen in the previous image, molecules are complex objects; they are a population of atoms and thermal agitation can strain their internal chemical bonds in three different ways: via rotation, bond length, and bond angle movements. These are all types of *internal degrees of freedom*. This makes molecules distinct from *monatomic* substances (consisting of individual atoms) like the noble gases helium and argon, which have only the three translational degrees of freedom. Kinetic energy is stored in molecules' internal degrees of freedom, which gives them an *internal temperature*. Even though these motions are called *internal*, the external portions of molecules still move—rather like the jiggling of a stationary water balloon. This permits the two-way exchange of kinetic energy between internal motions and translational motions with each molecular collision. Accordingly, as energy is removed from molecules, both their kinetic temperature (the temperature derived from the kinetic energy of translational motion) and their internal temperature simultaneously diminish in equal proportions. This phenomenon is described by the equipartition theorem, which states that for any bulk quantity of a substance in equilibrium, the kinetic energy of particle motion is evenly distributed among all the active (i.e. unfrozen) degrees of freedom available to the particles. Since the internal temperature of molecules are usually equal to their kinetic temperature, the distinction is usually of interest only in the detailed study of non-local thermodynamic equilibrium (LTE) phenomena such as combustion, the sublimation of solids, and the diffusion of hot gases in a partial vacuum.

The kinetic energy stored internally in molecules causes substances to contain more internal energy at any given temperature and to absorb additional internal energy for a given temperature increase. This is because any kinetic energy that is, at a given instant, bound in internal motions is not at that same instant contributing to the molecules' translational motions. This extra thermal energy simply increases the amount of energy a substance absorbs for a given temperature rise. This property is known as a substance's specific heat capacity.

Different molecules absorb different amounts of thermal energy for each incremental increase in temperature; that is, they have different specific heat capacities. High specific heat capacity arises, in part, because certain substances' molecules possess more internal degrees of freedom than others do. For instance, nitrogen, which is a diatomic molecule, has *five* active degrees of freedom at room temperature: the three comprising translational motion plus two rotational degrees of freedom internally. Since the two internal degrees of freedom are essentially unfrozen, in accordance with the equipartition theorem, nitrogen has five-thirds the specific heat capacity per mole (a specific number of molecules) as do the monatomic gases. Another example is gasoline. Gasoline can absorb a large amount of thermal energy per mole with only a modest temperature change because each molecule comprises an average of 21 atoms and therefore has many internal degrees of freedom. Even larger, more complex molecules can have dozens of internal degrees of freedom.

The Diffusion of Thermal Energy: Entropy, Phonons, and Mobile Conduction Electrons

Heat conduction is the diffusion of thermal energy from hot parts of a system to cold. A system can be either a single bulk entity or a plurality of discrete bulk entities. The term *bulk* in this context means a statistically significant quantity of particles (which can be a microscopic amount). Whenever thermal energy diffuses within an isolated system, temperature differences within the system decrease (and entropy increases).

One particular heat conduction mechanism occurs when translational motion, the particle motion underlying temperature, transfers momentum from particle to particle in collisions. In gases, these translational motions are of the nature shown above in *figure*. As can be seen, not only does momentum (heat) diffuse throughout the volume of the gas through serial collisions, but entire molecules or atoms can move forward into new territory, bringing their kinetic energy with them. Consequently, temperature differences equalize throughout gases very quickly—especially for light atoms or molecules; convection speeds this process even more.

Translational motion in *solids*, however, takes the form of *phonons*. Phonons are constrained, quantized wave packets that travel at a given substance's speed of sound. The manner in which phonons interact within a solid determines a variety of its properties, including its thermal conductivity. In electrically insulating solids, phonon-based heat conduction is *usually* inefficient and such solids are considered *thermal insulators* (such as glass, plastic, rubber, ceramic, and rock). This is because in solids, atoms and molecules are locked into place relative to their neighbors and are not free to roam.

Metals however, are not restricted to only phonon-based heat conduction. Thermal energy conducts through metals extraordinarily quickly because instead of direct molecule-to-molecule collisions, the vast majority of thermal energy is mediated via very light, mobile *conduction electrons*. This is why there is a near-perfect correlation between metals' thermal conductivity and their electrical conductivity. Conduction electrons imbue metals with their extraordinary conductivity because they are *delocalized* (i.e., not tied to a specific atom) and behave rather like a sort of quantum gas due to the effects of *zero-point energy*. Furthermore, electrons are relatively light with a rest mass only $\frac{1}{1836}$th that of a proton. This is about the same ratio as a .22 Short bullet (29 grains or 1.88 g) compared to the rifle that shoots it. As Isaac Newton wrote with his third law of motion,

Law #3: All forces occur in pairs, and these two forces are equal in magnitude and opposite in direction.

However, a bullet accelerates faster than a rifle given an equal force. Since kinetic energy increases as the square of velocity, nearly all the kinetic energy goes into the bullet, not the rifle, even though both experience the same force from the expanding propellant gases. In the same manner, because they are much less massive, thermal energy is readily borne by mobile conduction electrons. Additionally, because they're delocalized and *very* fast, kinetic thermal energy conducts extremely quickly through metals with abundant conduction electrons.

The Diffusion of Thermal Energy: Black-body Radiation

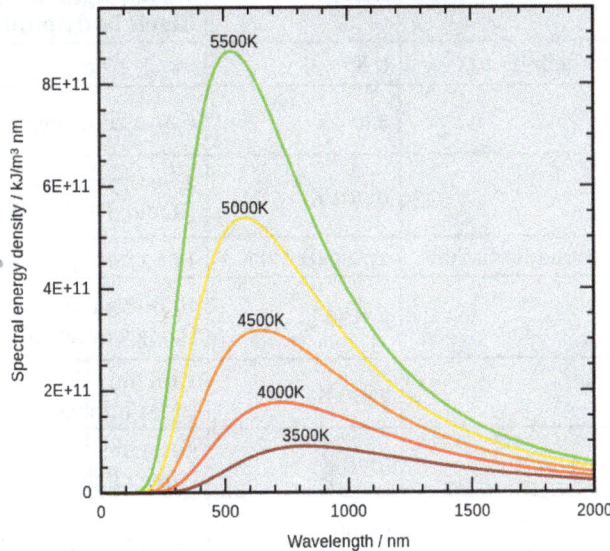

The spectrum of black-body radiation has the form of a Planck curve. A 5500 K
black-body has a peak emittance wavelength of 527 nm.

Thermal radiation is a byproduct of the collisions arising from various vibrational motions of atoms. These collisions cause the electrons of the atoms to emit thermal photons (known as black-body radiation). Photons are emitted anytime an electric charge is accelerated (as happens when electron clouds of two atoms collide). Even *individual molecules* with internal temperatures greater than absolute zero also emit black-body radiation from their atoms. In any bulk quantity of a substance at equilibrium, black-body photons are emitted across a range of wavelengths in a spectrum that has a bell curve-like shape called a Planck curve. The top of a Planck curve (the peak emittance wavelength) is located in a particular part of the electromagnetic spectrum depending on the temperature of the black-body. Substances at extreme cryogenic temperatures emit at long radio wavelengths whereas extremely hot temperatures produce short gamma rays.

Black-body radiation diffuses thermal energy throughout a substance as the photons are absorbed by neighboring atoms, transferring momentum in the process. Black-body photons also easily escape from a substance and can be absorbed by the ambient environment; kinetic energy is lost in the process.

As established by the Stefan–Boltzmann law, the intensity of black-body radiation increases as the fourth power of absolute temperature. Thus, a black-body at 824 K (just short of glowing dull red) emits *60 times* the radiant power as it does at 296 K (room temperature). This is why one can so easily feel the radiant heat from hot objects at a distance. At higher temperatures, such as those found in an incandescent lamp, black-body radiation can be the principal mechanism by which thermal energy escapes a system.

Table of Thermodynamic Temperatures

The full range of the thermodynamic temperature scale, from absolute zero to absolute hot, and some notable points between them are shown in the table below.

	kelvin	Peak emittance wavelength of black-body photons
Absolute zero (precisely by definition)	0 K	∞
Coldest measured temperature	450 pK	6,400 kilometers
One millikelvin (precisely by definition)	0.001 K	2.897 77 meters (Radio, FM band)
Cosmic Microwave Background Radiation	2.725 48(57) K	1.063 mm (peak wavelength)
Water's triple point (precisely by definition)	273.16 K	10,608.3 nm (Long wavelength I.R.)
Incandescent lamp[B]	2500 K	1160 nm (Near infrared)[C]
Sun's visible surface[C]	5778 K	501.5 nm (Green light)
Lightning bolt's channel	28,000 K	100 nm (Far Ultraviolet light)
Sun's core	16 MK	0.18 nm (X-rays)
Thermonuclear explosion (peak temperature)	350 MK	8.3×10^{-3} nm (Gamma rays)
Sandia National Labs' Z machine [D]	2 GK	1.4×10^{-3} nm (Gamma rays)
Core of a high–mass star on its last day	3 GK	1×10^{-3} nm (Gamma rays)
Merging binary neutron star system	350 GK	8×10^{-6} nm (Gamma rays)
Gamma-ray burst progenitors	1 TK	3×10^{-6} nm (Gamma rays)
Relativistic Heavy Ion Collider	1 TK	3×10^{-6} nm (Gamma rays)
CERN's proton vs. nucleus collisions	10 TK	3×10^{-7} nm (Gamma rays)
Universe 5.391×10^{-44} s after the Big Bang	1.417×10^{32} K	1.616×10^{-26} nm (Planck frequency)

A. The 2500 K value is approximate.

B. For a true blackbody (which tungsten filaments are not). Tungsten filaments' emissivity is greater at shorter wavelengths, which makes them appear whiter.

C. Effective photosphere temperature.

D. For a true blackbody (which the plasma was not). The Z machine's dominant emission originated from 40 MK electrons (soft x–ray emissions) within the plasma.

The Heat of Phase Changes

Ice and water: two phases of the same substance

The kinetic energy of particle motion is just one contributor to the total thermal energy in a substance; another is *phase transitions*, which are the potential energy of molecular bonds that can form in a substance as it cools (such as during condensing and freezing). The thermal energy required for a phase transition is called *latent heat*. This phenomenon may more easily be grasped by considering it in the reverse direction: latent heat is the energy required to *break* chemical bonds (such as during evaporation and melting). Almost everyone is familiar with the effects of phase transitions; for instance, steam at 100°C can cause severe burns much faster than the 100°C air from a hair dryer. This occurs because a large amount of latent heat is liberated as steam condenses into liquid water on the skin.

Even though thermal energy is liberated or absorbed during phase transitions, pure chemical elements, compounds, and eutectic alloys *exhibit no temperature change whatsoever* while they undergo them (*Fig* below). Consider one particular type of phase transition: melting. When a solid is melting, crystal lattice chemical bonds are being broken apart; the substance is transitioning from what is known as a *more ordered state* to a *less ordered state*. In the *figure below,* the melting of ice is shown within the lower left box heading from blue to green.

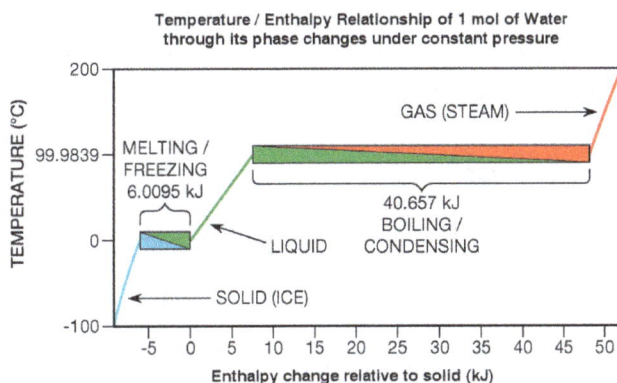

Water's temperature does not change during phase transitions as heat flows into or out of it. The total heat capacity of a mole of water in its liquid phase (the green line) is 7.5507 kJ.

At one specific thermodynamic point, the melting point (which is 0°C across a wide pressure range in the case of water), all the atoms or molecules are, on average, at the maximum energy threshold their chemical bonds can withstand without breaking away from the lattice. Chemical bonds are all-or-nothing forces: they either hold fast, or break; there is no in-between state. Consequently, when a substance is at its melting point, every joule of added thermal energy only breaks the bonds of a specific quantity of its atoms or molecules, converting them into a liquid of precisely the same temperature; no kinetic energy is added to translational motion (which is what gives substances their temperature). The effect is rather like popcorn: at a certain temperature, additional thermal energy can't make the kernels any hotter until the transition (popping) is complete. If the process is reversed (as in the freezing of a liquid), thermal energy must be removed from a substance.

As stated above, the thermal energy required for a phase transition is called *latent heat*. In the specific cases of melting and freezing, it's called *enthalpy of fusion* or *heat of fusion*. If the molecular bonds in a crystal lattice are strong, the heat of fusion can be relatively great, typically in the range of 6 to 30 kJ per mole for water and most of the metallic elements. If the substance is one of the monatomic gases, (which have little tendency to form molecular bonds) the heat of fusion is more modest, ranging from 0.021 to 2.3 kJ per mole. Relatively speaking, phase transitions can be truly energetic events. To completely melt ice at 0°C into water at 0°C, one must add roughly 80 times the thermal energy as is required to increase the temperature of the same mass of liquid water by one degree Celsius. The metals' ratios are even greater, typically in the range of 400 to 1200 times. And the phase transition of boiling is much more energetic than freezing. For instance, the energy required to completely boil or vaporize water (what is known as *enthalpy of vaporization*) is roughly *540 times* that required for a one-degree increase.

Water's sizable enthalpy of vaporization is why one's skin can be burned so quickly as steam condenses on it (heading from red to green in *Fig.* above). In the opposite direction, this is why one's skin feels cool as liquid water on it evaporates (a process that occurs at a sub-ambient wet-bulb temperature that is dependent on relative humidity). Water's highly energetic enthalpy of vaporization is also an important factor underlying why *solar pool covers* (floating, insulated blankets that cover swimming pools when not in use) are so effective at reducing heating costs: they prevent evaporation. For instance, the evaporation of just 20 mm of water from a 1.29-meter-deep pool chills its water 8.4 degrees Celsius (15.1°F).

Internal Energy

The total energy of all particle motion translational and internal, including that of conduction electrons, plus the potential energy of phase changes, plus zero-point energy comprise the *internal energy* of a substance.

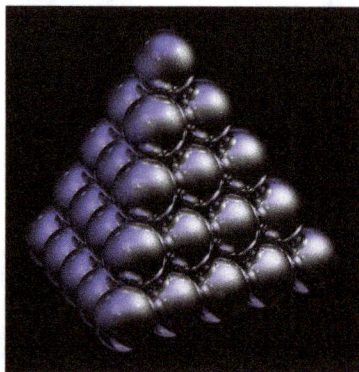

When many of the chemical elements, such as the noble gases and platinum-group metals, freeze to a solid — the most ordered state of matter — their crystal structures have a *closest-packed arrangement*. This yields the greatest possible packing density and the lowest energy state.

Internal Energy at Absolute Zero

As a substance cools, different forms of internal energy and their related effects simultaneously decrease in magnitude: the latent heat of available phase transitions is liberated as a substance changes from a less ordered state to a more ordered state; the translational motions of atoms and molecules diminish (their kinetic temperature decreases); the internal motions of molecules diminish (their internal temperature decreases); conduction electrons (if the substance is an electrical conductor) travel *somewhat* slower; and black-body radiation's peak emittance wavelength increases (the photons' energy decreases). When the particles of a substance are as close as possible to complete rest and retain only ZPE-induced quantum mechanical motion, the substance is at the temperature of absolute zero ($T=0$).

Note that whereas absolute zero is the point of zero thermodynamic temperature and is also the point at which the particle constituents of matter have minimal motion, absolute zero is not necessarily the point at which a substance contains zero thermal energy; one must be very precise with what one means by *internal energy*. Often, all the phase changes that *can* occur in a substance, *will* have occurred by the time it reaches absolute zero. However, this is not always the case. Notably, $T=0$ helium remains liquid at room pressure and must be under a pressure of at least 25 bar (2.5 MPa) to crystallize. This is because helium's heat of fusion (the energy required to melt helium ice) is so low (only 21 joules per mole) that the motion-inducing effect of zero-point energy is sufficient to prevent it from freezing at lower pressures. Only if under at least 25 bar (2.5 MPa) of pressure will this latent thermal energy be liberated as helium freezes while approaching absolute zero. A further complication is that many solids change their crystal structure to more compact arrangements at extremely high pressures (up to millions of bars, or hundreds of gigapascals). These are known as *solid-solid phase transitions* wherein latent heat is liberated as a crystal lattice changes to a more thermodynamically favorable, compact one.

The above complexities make for rather cumbersome blanket statements regarding the internal energy in $T=0$ substances. Regardless of pressure though, what *can* be said is that at absolute zero, all solids with a lowest-energy crystal lattice such those with a *closest-packed arrangement* (*Fig.* above) contain minimal internal energy, retaining only that due to the ever-present background of zero-point energy. One can also say that for a given substance at constant pressure, absolute zero is the point of lowest *enthalpy* (a measure of work potential that takes internal energy, pressure, and volume into consideration). Lastly, it is always true to say that all $T=0$ substances contain zero kinetic thermal energy.

Practical Applications for Thermodynamic Temperature

Helium-4, is a superfluid at or below 2.17 kelvins, (2.17 Celsius degrees above absolute zero)

Thermodynamic temperature is useful not only for scientists, it can also be useful for lay-people in many disciplines involving gases. By expressing variables in absolute terms and applying Gay–Lussac's law of temperature/pressure proportionality, solutions to everyday problems are straightforward; for instance, calculating how a temperature change affects the pressure inside an automobile tire. If the tire has a relatively cold pressure of 200 kPa-gage , then in absolute terms (relative to a vacuum), its pressure is 300 kPa-absolute. Room temperature ("cold" in tire terms) is 296 K. If the tire pressure is 20°C hotter (20 kelvins), the solution is calculated as $316 \, \text{K}/_{296 \, \text{K}} = 6.8\%$ greater thermodynamic temperature *and* absolute pressure; that is, a pressure of 320 kPa-absolute, which is 220 kPa-gage.

Definition of Thermodynamic Temperature

The thermodynamic temperature is defined by the second law of thermodynamics and its consequences. The thermodynamic temperature can be shown to have special properties, and in particular can be seen to be uniquely defined (up to some constant multiplicative factor) by considering the efficiency of idealized heat engines. Thus the *ratio* T_2/T_1 of two temperatures T_1 and T_2 is the same in all absolute scales.

Strictly speaking, the temperature of a system is well-defined only if it is at thermal equilibrium. From a microscopic viewpoint, a material is at thermal equilibrium if the quantity of heat between its individual particles cancel out. There are many possible scales of temperature, derived from a variety of observations of physical phenomena.

Loosely stated, temperature differences dictate the direction of heat between two systems such that their combined energy is maximally distributed among their lowest possible states. We call this distribution "entropy". To better understand the relationship between temperature and entropy, consider the relationship between heat, work and temperature illustrated in the Carnot heat engine. The engine converts heat into work by directing a temperature gradient between a higher temperature heat source, T_H, and a lower temperature heat sync, T_C, through a gas filled piston. The work done per cycle is equal to the difference between the heat supplied to the engine by T_H, q_H, and the heat supplied to T_C by the engine, q_C. The efficiency of the engine is the work divided by the heat put into the system or

$$\text{Efficiency} = \frac{w_{cy}}{q_H} = \frac{q_H - q_C}{q_H} = 1 - \frac{q_C}{q_H} \qquad (1)$$

where w_{cy} is the work done per cycle. Thus the efficiency depends only on q_C/q_H.

Carnot's theorem states that all reversible engines operating between the same heat reservoirs are equally efficient. Thus, any reversible heat engine operating between temperatures T_1 and T_2 must have the same efficiency, that is to say, the efficiency is the function of only temperatures

$$\frac{q_C}{q_H} = f(T_H, T_C) \qquad (2).$$

In addition, a reversible heat engine operating between temperatures T_1 and T_3 must have the same efficiency as one consisting of two cycles, one between T_1 and another (intermediate) temperature T_2, and the second between T_2 and T_3. If this were not the case, then energy (in the form of Q) will be wasted or gained, resulting in different overall efficiencies every time a cycle is split into component cycles; clearly a cycle can be composed of any number of smaller cycles.

With this understanding of Q_1, Q_2 and Q_3, we note also that mathematically,

$$f(T_1,T_3) = \frac{q_3}{q_1} = \frac{q_2 q_3}{q_1 q_2} = f(T_1,T_2)f(T_2,T_3).$$

But the first function is *NOT* a function of T_2, therefore the product of the final two functions *MUST* result in the removal of T_2 as a variable. The only way is therefore to define the function f as follows:

$$f(T_1,T_2) = \frac{g(T_2)}{g(T_1)}.$$

and

$$f(T_2,T_3) = \frac{g(T_3)}{g(T_2)}.$$

so that

$$f(T_1,T_3) = \frac{g(T_3)}{g(T_1)} = \frac{q_3}{q_1}.$$

i.e. The ratio of heat exchanged is a function of the respective temperatures at which they occur. We can choose any monotonic function for our $g(T)$; it is a matter of convenience and convention that we choose $g(T) = T$. Choosing then *one* fixed reference temperature (i.e. triple point of water), we establish the thermodynamic temperature scale.

It is to be noted that such a definition coincides with that of the ideal gas derivation; also it is this *definition* of the thermodynamic temperature that enables us to represent the Carnot efficiency in terms of T_H and T_C, and hence derive that the (complete) Carnot cycle is isentropic:

$$\frac{q_C}{q_H} = f(T_H,T_C) = \frac{T_C}{T_H}. \qquad (3).$$

Substituting this back into our first formula for efficiency yields a relationship in terms of temperature:

$$\text{Efficiency} = 1 - \frac{q_C}{q_H} = 1 - \frac{T_C}{T_H} \qquad (4).$$

Notice that for $T_C = 0$ the efficiency is 100% and that efficiency becomes greater than 100% for $T_C < 0$, which cases are unrealistic. Subtracting the right hand side of Equation 4 from the middle portion and rearranging gives

$$\frac{q_H}{T_H} - \frac{q_C}{T_C} = 0,$$

where the negative sign indicates heat ejected from the system. The generalization of this equation is Clausius theorem, which suggests the existence of a state function S (i.e., a function which depends only on the state of the system, not on how it reached that state) defined (up to an additive constant) by

$$dS = \frac{dq_{rev}}{T} \qquad (5),$$

where the subscript indicates heat transfer in a reversible process. The function S corresponds to the entropy of the system, mentioned previously, and the change of S around any cycle is zero (as is necessary for any state function). Equation 5 can be rearranged to get an alternative definition for temperature in terms of entropy and heat (to avoid logic loop, we should first define entropy through statistical mechanics):

$$T = \frac{dq_{rev}}{dS}.$$

For a system in which the entropy S is a function $S(E)$ of its energy E, the thermodynamic temperature T is therefore given by

$$\frac{1}{T} = \frac{dS}{dE},$$

so that the reciprocal of the thermodynamic temperature is the rate of increase of entropy with energy.

References

- Junling Huang & Michael B. McElroy (2014). "Contributions of the Hadley and Ferrel Circulations to the Energetics of the Atmosphere over the Past 32 Years". Journal of Climate. 27 (7): 2656–2666. Bibcode:2014JCli...27.2656H. doi:10.1175/jcli-d-13-00538.1

- Yikne Deng (2005). Ancient Chinese Inventions. Chinese International Press. pp. 112–13. ISBN 978-7-5085-0837-5. Retrieved 2009-06-18

- National Center for Atmospheric Research (2008). "Hail". University Corporation for Atmospheric Research. Retrieved 2009-07-18

- Junling Huang & Michael B. McElroy (2015). "Thermodynamic disequilibrium of the atmosphere in the context of global warming". Climate Dynamics. Bibcode:2015ClDy..tmp...98H. doi:10.1007/s00382-015-2553-x

- Chris C. Mooney (2007). Storm world: hurricanes, politics, and the battle over global warming. Houghton Mifflin Harcourt. p. 20. ISBN 978-0-15-101287-9. Retrieved 2009-08-31

- Jacque Marshall (2000-04-10). "Hail Fact Sheet". University Corporation for Atmospheric Research. Retrieved 2009-07-15

- Renno, Nilton O. (August 2008). "A thermodynamically general theory for convective vortices" (PDF). Tellus A. 60 (4): 688–99. Bibcode:2008TellA..60..688R. doi:10.1111/j.1600-0870.2008.00331.x

- Douglas G. Hahn and Syukuro Manabe (1975). "The Role of Mountains in the South Asian Monsoon Circulation". Journal of Atmospheric Sciences. 32 (8): 1515–1541. Bibcode:1975JAtS...32.1515H. ISSN 1520-0469. doi:10.1175/1520-0469(1975)032<1515:TROMIT>2.0.CO;2

- Frank W. Gallagher, III. (October 2000). "Distant Green Thunderstorms - Frazer's Theory Revisited". Journal of Applied Meteorology. American Meteorological Society. 39 (10): 1754. Bibcode:2000JApMe..39.1754G. doi:10.1175/1520-0450-39.10.1754. Retrieved 2011-01-20

- Michael H. Mogil (2007). Extreme Weather. New York: Black Dog & Leventhal Publisher. pp. 210–211. ISBN 978-1-57912-743-5

- Edwards, Roger (2006-04-04). "The Online Tornado FAQ". Storm Prediction Center. Archived from the original on September 30, 2006. Retrieved 2006-09-08

- M. W. Moncrieff; M.J. Miller (1976). "The dynamics and simulation of tropical cumulonimbus and squall lines" (abstract). Q. J. R. Meteorol. Soc. 120 (432): 373–94. Bibcode:1976QJRMS.102..373M. doi:10.1002/qj.49710243208

- Stephan P. Nelson (August 1983). "The Influence of Storm Flow Struce on Hail Growth". Journal of Atmospheric Sciences. 40 (8): 1965–1983. Bibcode:1983JAtS...40.1965N. ISSN 1520-0469. doi:10.1175/1520-0469(1983)040<1965:TIOSFS>2.0.CO;2

- John E. Oliver (2005-08-23). Encyclopedia of world climatology. Springer. p. 449. ISBN 978-1-4020-3264-6. Retrieved 2012-02-24

- "Doppler On Wheels". Center for Severe Weather Research. 2006. Archived from the original on 5 February 2007. Retrieved 2006-12-29.

- Michael Vollmer (March 2009). "Mirrors in the air: mirages in nature and in the laboratory". Physics Education. IOP Publishing Limited. 44 (2): 167. Bibcode:2009PhyEd..44..165V. doi:10.1088/0031-9120/44/2/008

- Edward Aguado & James E. Burt (2007). Understanding weather and climate. Pearson Prentice Hall. pp. 416–418. ISBN 978-0-13-149696-5

- "Hallam Nebraska Tornado". Omaha/Valley, NE Weather Forecast Office. 2005-10-02. Archived from the original on 4 October 2006. Retrieved 2006-09-08

- M. W. Moncrieff, M.J. Miller (1976). "The dynamics and simulation of tropical cumulonimbus and squall lines". Q. J. R. Meteorol. Soc. 120 (432): 373–94. Bibcode:1976QJRMS.102..373M. doi:10.1002/qj.49710243208

7

Atmospheric Temperature Measurement

Atmospheric temperature is measured by using various tools, techniques and concepts. The scale of temperature quantitatively measures temperature through the ideal gas scale and the International temperature scale of 1990. The chapter strategically encompasses and incorporates the major components and key concepts of atmospheric temperature measurement, providing a complete understanding.

Temperature Measurement

Daniel Gabriel Fahrenheit, the originator of the era of precision thermometry.
He invented the mercury thermometer (first practical, accurate thermometer) and Fahrenheit
scale (first standardized temperature scale to be widely used).

Temperature measurement (or thermometry) describes the process of measuring a current local temperature for immediate or later evaluation. Datasets consisting of repeated standardized measurements can be used to assess temperature trends.

History

Attempts at standardized temperature measurement prior to the 17th century were crude at best. For instance in 170 AD, physician Claudius Galenus mixed equal portions of ice and boiling water to create a "neutral" temperature standard. The modern scientific field has its origins in the works by Florentine scientists in the 1600s including Galileo constructing devices able to measure relative change in temperature, but subject also to confounding with atmospheric pressure changes. These early devices were called thermoscopes. The first sealed thermometer was constructed in 1641 by the Grand Duke of Toscani, Ferdinand II. The development of today's thermometers and temperature scales began in the early 18th century, when Gabriel Fahrenheit produced a mercury thermometer and scale, both developed by Ole Christensen Rømer. Fahrenheit's scale is still in use, alongside the Celsius and Kelvin scales.

Technologies

Many methods have been developed for measuring temperature. Most of these rely on measuring some physical property of a working material that varies with temperature. One of the most common devices for measuring temperature is the glass thermometer. This consists of a glass tube filled with mercury or some other liquid, which acts as the working fluid. Temperature increase causes the fluid to expand, so the temperature can be determined by measuring the volume of the fluid. Such thermometers are usually calibrated so that one can read the temperature simply by observing the level of the fluid in the thermometer. Another type of thermometer that is not really used much in practice, but is important from a theoretical standpoint, is the gas thermometer.

Other important devices for measuring temperature include:

- Thermocouples

- Thermistors

- Resistance temperature detector (RTD)

- Pyrometer

- Langmuir probes (for electron temperature of a plasma)

- Infrared

- Other thermometers

One must be careful when measuring temperature to ensure that the measuring instrument (thermometer, thermocouple, etc.) is really the same temperature as the material that is being measured. Under some conditions heat from the measuring instrument can cause a temperature gradient, so the measured temperature is different from the actual temperature of the system. In such a case the measured temperature will vary not only with the temperature of the system, but also with the heat transfer properties of the system. An extreme case of this effect gives rise to the wind chill factor, where the weather feels colder under windy conditions than calm conditions even though the temperature is the same. What is happening is that the wind increases the rate of heat transfer from the body, resulting in a larger reduction in body temperature for the same ambient temperature.

The theoretical basis for thermometers is the zeroth law of thermodynamics which postulates that if you have three bodies, A, B and C, if A and B are at the same temperature, and B and C are at the same temperature then A and C are at the same temperature. B, of course, is the thermometer.

The practical basis of thermometry is the existence of triple point cells. Triple points are conditions of pressure, volume and temperature such that three phases are simultaneously present, for example solid, vapor and liquid. For a single component there are no degrees of freedom at a triple point and any change in the three variables results in one or more of the phases vanishing from the cell. Therefore, triple point cells can be used as universal references for temperature and pressure.

Under some conditions it becomes possible to measure temperature by a direct use of the Planck's law of black-body radiation. For example, the cosmic microwave background temperature has been measured from the spectrum of photons observed by satellite observations such as the WMAP. In the study of the quark–gluon plasma through heavy-ion collisions, single particle spectra sometimes serve as a thermometer.

Non-invasive Thermometry

During recent decades, many thermometric techniques have been developed. The most promising and widespread non-invasive thermometric techniques are based on the analysis of magnetic resonance images, computerised tomography images and echotomography images. These techniques allow monitoring temperature within tissues without introducing a sensing element.

Surface Air Temperature

The temperature of the air near the surface of the Earth is measured at meteorological observatories and weather stations, usually using thermometers placed in a Stevenson screen, a standardized well-ventilated white-painted instrument shelter. The thermometers should be positioned 1.25–2 m above the ground. Details of this setup are defined by the World Meteorological Organization (WMO).

A true daily mean could be obtained from a continuously-recording thermograph. Commonly it is approximated by the mean of discrete readings (e.g. 24 hourly readings, four 6-hourly readings, etc.) or by the mean of the daily minimum and maximum readings (though the latter can result in mean temperatures up to 1°C cooler or warmer than the true mean, depending on the time of observation).

The world's average surface air temperature is about 14°C. For information on temperature changes relevant to climate change or Earth's geologic past.

Comparison of Temperature Scales

Comparison of temperature scales								
Comment	Kelvin K	Celsius °C	Fahrenheit °F	Rankine °Ra (°R)	Delisle °D [1]	Newton °N	Réaumur °R (°Ré, °Re) [1]	Rømer °Rø (°R) [1]
Absolute zero	0	−273.15	−459.67	0	559.725	−90.14	−218.52	−135.90
Lowest recorded natural temperature on Earth (Vostok, Antarctica - 21 July 1983)	184	−89	−128	331	284	−29	−71	−39
Celsius / Fahrenheit's "crossover" temperature	233.15	−40	−40	419.67	210	−13.2	−32	−13.5
Fahrenheit's ice/salt mixture	255.37	−17.78	0	459.67	176.67	−5.87	−14.22	−1.83
Water freezes (at standard pressure)	273.15	0	32	491.67	150	0	0	7.5

Average surface temperature on Earth	287	14	57	517	129	4.6	12	15.4
Average human body temperature [2]	310.0 ±0.7	36.8 ±0.7	98.2 ±1.3	557.9 ±1.3	94.8 ±1.1	12.1 ±0.2	29.4 ±0.6	26.8 ±0.4
Highest recorded surface temperature on Earth (Furnace Creek Ranch, USA - 10 July 1913)	329.8	56.7	134	593.7	65.0	18.7	45.3	37.3
Water boils (at standard pressure)	373.15	100	212	672	0	33	80	60
Gas flame	~1773	~1500	~2732					
Titanium melts	1941	1668	3034	3494	−2352	550	1334	883
The surface of the Sun	5800	5526	9980	10440	−8140	1823	4421	2909

[1] The temperature scale is in disuse, and of mere historical interest.

[2] Normal human body temperature is 36.8 ±0.7°C, or 98.2 ±1.3°F. The commonly given value 98.6°F is simply the exact conversion of the nineteenth-century German standard of 37°C. Since it does not list an acceptable range, it could therefore be said to have excess (invalid) precision.Some numbers in this table have been rounded off.

Standards

The American Society of Mechanical Engineers (ASME) has developed two separate and distinct standards on temperature Measurement, B40.200 and PTC 19.3. B40.200 provides guidelines for bimetallic-actuated, filled-system, and liquid-in-glass thermometers. It also provides guidelines for thermowells. PTC 19.3 provides guidelines for temperature measurement related to Performance Test Codes with particular emphasis on basic sources of measurement errors and techniques for coping with them.

US (ASME) Standards

- B40.200-2008: Thermometers, Direct Reading and Remotes Reading.
- PTC 19.3-1974(R2004): Performance test code for temperature measurement.

Scale of Temperature

Scale of temperature is a way to measure temperature quantitatively. Empirical scales measure the quantity of heat in a system in relation to a fixed parameter, a thermometer. They are not absolute measures, that is why scales vary. Absolute temperature is thermodynamic temperature because it is directly related to thermodynamics. It is the Zeroth Law of Thermodynamics that leads to a formal definition of thermodynamic temperature.

Formal Description

According to the zeroth law of thermodynamics, being in thermal equilibrium is an equivalence relation. Thus all thermal systems may be divided into a quotient set by this equivalence relation, denoted below as M. Assume the set M has the cardinality of c, then one can construct an injective function $f: M \rightarrow R$, by which every thermal system will have a number associated with it such that when and only when two thermal systems have the same such value, they will be in thermal equilibrium. This is clearly the property of temperature, and the specific way of assigning numerical values as temperature is called a *scale of temperature*. In practical terms, a temperature scale is always based on usually a single physical property of a simple thermodynamic system, called a thermometer, that defines a scaling function mapping the temperature to the measurable thermometric parameter. Such temperature scales that are purely based on measurement are called *empirical temperature scales*.

The second law of thermodynamics provides a fundamental, natural definition of thermodynamic temperature starting with a null point of absolute zero. A scale for thermodynamic temperature is established similarly to the empirical temperature scales, however, needing only one additional fixing point.

Empirical Scales

Empirical scales are based on the measurement of physical parameters that express the property of interest to be measured through some formal, most commonly a simple linear, functional relationship. For the measurement of temperature, the formal definition of thermal equilibrium in terms of the thermodynamic coordinate spaces of thermodynamic systems, expressed in the zeroth law of thermodynamics, provides the framework to measure temperature.

All temperature scales, including the modern thermodynamic temperature scale used in the International System of Units, are calibrated according to thermal properties of a particular substance or device. Typically, this is established by fixing two well-defined temperature points and defining temperature increments via a linear function of the response of the thermometric device. For example, both the old Celsius scale and Fahrenheit scale were originally based on the linear expansion of a narrow mercury column within a limited range of temperature, each using different reference points and scale increments.

Different empirical scales may not be compatible with each other, except for small regions of temperature overlap. If an alcohol thermometer and a mercury thermometer have same two fixed points, namely the freezing and boiling point of water, their reading will not agree with each other except at the fixed points, as the linear 1:1 relationship of expansion between any two thermometric substances may not be guaranteed.

Empirical temperature scales are not reflective of the fundamental, microscopic laws of matter. Temperature is a universal attribute of matter, yet empirical scales map a narrow range onto a scale that is known to have a useful functional form for a particular application. Thus, their range is limited. The working material only exists in a form under certain circumstances, beyond which it no longer can serve as a scale. For example, mercury freezes below 234.32 K, so temperature lower than that cannot be measured in a scale based on mercury. Even ITS-90, which interpolates among different ranges of temperature, has only a range of 0.65 K to approximately 1358 K (−272.5°C to 1085°C).

Ideal Gas Scale

When pressure approaches zero, all real gas will behave like ideal gas, that is, pV of a mole of gas relying only on temperature. Therefore, we can design a scale with pV as its argument. Of course any bijective function will do, but for convenience's sake linear function is the best. Therefore, we define it as

$$T = \frac{1}{nR} \lim_{p \to 0} pV.$$

The ideal gas scale is in some sense a "mixed" scale. It relies on the universal properties of gas, a big advance from just a particular substance. But still it is empirical since it puts gas at a special position and thus has limited applicability—at some point no gas can exist. One distinguishing characteristic of ideal gas scale, however, is that it precisely equals thermodynamical scale when it is well defined.

International Temperature Scale of 1990

ITS-90 is designed to represent the thermodynamic temperature scale (referencing absolute zero) as closely as possible throughout its range. Many different thermometer designs are required to cover the entire range. These include helium vapor pressure thermometers, helium gas thermometers, standard platinum resistance thermometers (known as SPRTs, PRTs or Platinum RTDs) and monochromatic radiation thermometers.

Although the Kelvin and Celsius scales are defined using absolute zero (0 K) and the triple point of water (273.16 K and 0.01°C), it is impractical to use this definition at temperatures that are very different from the triple point of water. Accordingly, ITS−90 uses numerous defined points, all of which are based on various thermodynamic equilibrium states of fourteen pure chemical elements and one compound (water). Most of the defined points are based on a phase transition; specifically the melting/freezing point of a pure chemical element. However, the deepest cryogenic points are based exclusively on the vapor pressure/temperature relationship of helium and its isotopes whereas the remainder of its cold points (those less than room temperature) are based on triple points. Examples of other defining points are the triple point of hydrogen (−259.3467°C) and the freezing point of aluminum (660.323°C).

Thermometers calibrated per ITS−90 use complex mathematical formulas to interpolate between its defined points. ITS−90 specifies rigorous control over variables to ensure reproducibility from lab to lab. For instance, the small effect that atmospheric pressure has upon the various melting points is compensated for (an effect that typically amounts to no more than half a millikelvin across the different altitudes and barometric pressures likely to be encountered). The standard even compensates for the pressure effect due to how deeply the temperature probe is immersed into the sample. ITS–90 also draws a distinction between "freezing" and "melting" points. The distinction depends on whether heat is going *into* (melting) or *out of* (freezing) the sample when the measurement is made. Only gallium is measured while melting, all the other metals are measured while the samples are freezing.

There are often small differences between measurements calibrated per ITS–90 and thermodynamic temperature. For instance, precise measurements show that the boiling point of VSMOW water under one standard atmosphere of pressure is actually 373.1339 K (99.9839°C) when adhering *strictly* to the two-point definition of thermodynamic temperature. When calibrated to ITS–90, where one must interpolate between the defining points of gallium and indium, the boiling point of VSMOW water is about 10 mK less, about 99.974°C. The virtue of ITS–90 is that another lab in another part of the world will measure the very same temperature with ease due to the advantages of a comprehensive international calibration standard featuring many conveniently spaced, reproducible, defining points spanning a wide range of temperatures.

Celsius Scale

Celsius (known until 1948 as centigrade) is a temperature scale that is named after the Swedish astronomer Anders Celsius (1701–1744), who developed a similar temperature scale two years before his death. The degree Celsius (°C) can refer to a specific temperature on the Celsius scale as well as a unit to indicate a temperature *interval* (a difference between two temperatures or an uncertainty).

From 1744 until 1954, 0°C was defined as the freezing point of water and 100°C was defined as the boiling point of water, both at a pressure of one standard atmosphere. Although these defining correlations are commonly taught in schools today, by international agreement the unit "degree Celsius" and the Celsius scale are currently defined by two different points: absolute zero, and the triple point of VSMOW (specially prepared water). This definition also precisely relates the Celsius scale to the Kelvin scale, which defines the SI base unit of thermodynamic temperature (symbol: K). Absolute zero, the hypothetical but unattainable temperature at which matter exhibits zero entropy, is defined as being precisely 0 K *and* −273.15°C. The temperature value of the triple point of water is defined as being precisely 273.16 K *and* 0.01°C.

This definition fixes the magnitude of both the degree Celsius and the kelvin as precisely 1 part in 273.16 parts, the difference between absolute zero and the triple point of water. Thus, it sets the magnitude of one degree Celsius and that of one kelvin as exactly the same. Additionally, it establishes the difference between the two scales' null points as being precisely 273.15 degrees Celsius (−273.15°C = 0 K and 0°C = 273.15 K).

Thermodynamic Scale

Thermodynamic scale differs from empirical scales in that it is absolute. It is based on the fundamental laws of thermodynamics or statistical mechanics instead of some arbitrary chosen working material. Besides it covers full range of temperature and has simple relation with microscopic quantities like the average kinetic energy of particles. In experiments ITS-90 is used to approximate thermodynamic scale due to simpler realization.

Definition

Lord Kelvin devised the thermodynamic scale based on the efficiency of heat engines as shown below:

The efficiency of an engine is the work divided by the heat introduced to the system or

$$\eta = \frac{w_{cy}}{q_H} = \frac{q_H - q_C}{q_H} = 1 - \frac{q_C}{q_H} \qquad (1)$$

where w_{cy} is the work done per cycle. Thus, the efficiency depends only on q_C/q_H.

Because of Carnot theorem, any reversible heat engine operating between temperatures T_1 and T_2 must have the same efficiency, meaning, the efficiency is the function of the temperatures only:

$$\frac{q_C}{q_H} = f(T_H, T_C) \qquad (2).$$

In addition, a reversible heat engine operating between temperatures T_1 and T_3 must have the same efficiency as one consisting of two cycles, one between T_1 and another (intermediate) temperature T_2, and the second between T_2 and T_3. This can only be the case if

$$f(T_1, T_3) = \frac{q_3}{q_1} = \frac{q_2 q_3}{q_1 q_2} = f(T_1, T_2) f(T_2, T_3).$$

Specializing to the case that T_1 is a fixed reference temperature: the temperature of the triple point of water. Then for any T_2 and T_3,

$$f(T_2, T_3) = \frac{f(T_1, T_3)}{f(T_1, T_2)} = \frac{273.16 \cdot f(T_1, T_3)}{273.16 \cdot f(T_1, T_2)}.$$

Therefore, if thermodynamic temperature is defined by

$$T = 273.16 \cdot f(T_1, T)$$

then the function f, viewed as a function of thermodynamic temperature, is

$$f(T_2, T_3) = \frac{T_3}{T_2},$$

and the reference temperature T_1 has the value 273.16. (Of course any reference temperature and any positive numerical value could be used—the choice here corresponds to the Kelvin scale.)

Equality to Ideal Gas Scale

It follows immediately that

$$\frac{q_C}{q_H} = f(T_H, T_C) = \frac{T_C}{T_H}. \qquad (3).$$

Substituting Equation 3 back into Equation 1 gives a relationship for the efficiency in terms of temperature:

$$\eta = 1 - \frac{q_C}{q_H} = 1 - \frac{T_C}{T_H} \quad (4).$$

This is identical to the efficiency formula for Carnot cycle, which effectively employs the ideal gas scale. This means that the two scales equal numerically at every point.

Conversion Table Between Different Temperature Scales

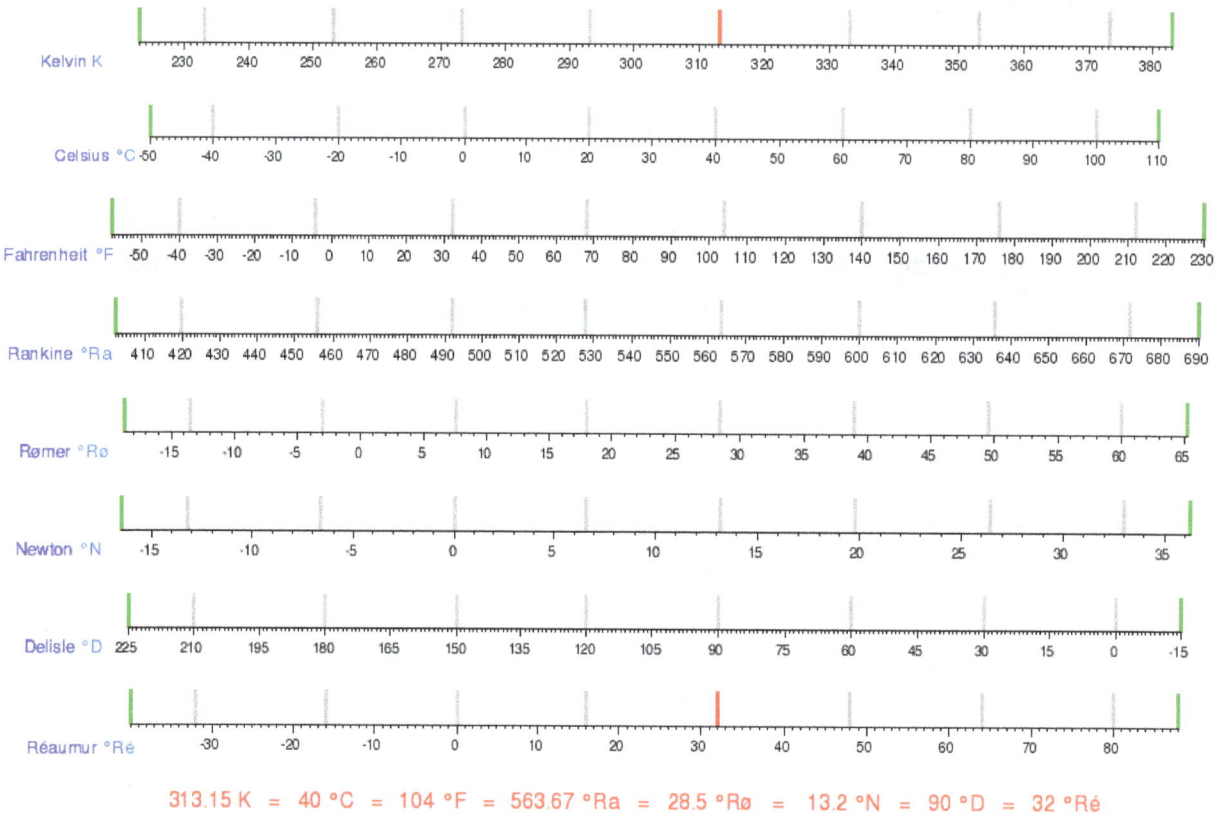

313.15 K = 40 °C = 104 °F = 563.67 °Ra = 28.5 °Rø = 13.2 °N = 90 °D = 32 °Ré

International Temperature Scale of 1990

The International Temperature Scale of 1990 (ITS-90) published by the Consultative Committee for Thermometry (CCT) of the International Committee for Weights and Measures (CIPM) is an equipment calibration standard for making measurements on the Kelvin and Celsius temperature scales. ITS–90 is an approximation of the thermodynamic temperature scale that facilitates the comparability and compatibility of temperature measurements internationally. It specifies fourteen calibration points ranging from 0.65±0 K to 1357.77±0 K (-272.50±0°C to 1084.62±0°C) and is subdivided into multiple temperature ranges which overlap in some instances. ITS-90 is the latest (as of 2014) of a series of International Temperature Scales adopted by CIPM since 1927. Adopted at the 1989 General Conference on Weights and Measures, it supersedes the International Practical Temperature Scale of 1968 (amended edition of 1975) and the 1976 "Provisional 0.5

K to 30 K Temperature Scale". CCT has also adopted a *mise en pratique* (practical instructions) in 2011. The lowest temperature covered by ITS-90 is 0.65 K. In 2000, the temperature scale was extended further, to 0.9 mK, by the adoption of a supplemental scale, known as the Provisional Low Temperature Scale of 2000 (PLTS-2000).

Details

CCT ITS-90 is designed to represent the thermodynamic (absolute) temperature scale (referencing absolute zero) as closely as possible throughout its range. Many different thermometer designs are required to cover the entire range. These include helium vapor pressure thermometers, helium gas thermometers, standard platinum resistance thermometers (known as SPRTs, PRTs or Platinum RTDs) and monochromatic radiation thermometers.

Although the Kelvin and Celsius scales are defined using absolute zero (0 K) and the triple point of water (273.16 K and 0.01°C), it is impractical to use this definition at temperatures that are very different from the triple point of water. Accordingly, ITS–90 uses numerous defined points, all of which are based on various thermodynamic equilibrium states of fourteen pure chemical elements and one compound (water). Most of the defined points are based on a phase transition; specifically the melting/freezing point of a pure chemical element. However, the deepest cryogenic points are based exclusively on the vapor pressure/temperature relationship of helium and its isotopes whereas the remainder of its cold points (those less than room temperature) are based on triple points. Examples of other defining points are the triple point of hydrogen (−259.3467°C) and the freezing point of aluminum (660.323°C).

Thermometers calibrated per ITS–90 use complex mathematical formulas to interpolate between its defined points. ITS–90 specifies rigorous control over variables to ensure reproducibility from lab to lab. For instance, the small effect that atmospheric pressure has upon the various melting points is compensated for (an effect that typically amounts to no more than half a millikelvin across the different altitudes and barometric pressures likely to be encountered). The standard even compensates for the pressure effect due to how deeply the temperature probe is immersed into the sample. ITS–90 also draws a distinction between "freezing" and "melting" points. The distinction depends on whether heat is going *into* (melting) or *out of* (freezing) the sample when the measurement is made. Only gallium is measured while melting, all the other metals are measured while the samples are freezing.

A practical effect of ITS–90 is the triple points and the freezing/melting points of its thirteen chemical elements are precisely known for all temperature measurements calibrated per ITS–90 since these thirteen values are fixed by its definition. Only the triple point of Vienna Standard Mean Ocean Water (VSMOW) is known with absolute precision—regardless of the calibration standard employed—because the very definitions of both the Kelvin and Celsius scales are fixed by international agreement based, in part, on this point.

Limitations

There are often small differences between measurements calibrated per ITS–90 and thermodynamic temperature. For instance, precise measurements show that the boiling point of VSMOW water under one standard atmosphere of pressure is actually 373.1339 K (99.9839°C) when adhering *strictly* to the two-point definition of thermodynamic temperature. When calibrated to ITS–90, where one must interpolate between the defining points of gallium and indium, the boiling point of VSMOW water is about 10 mK less, about 99.974°C. The virtue of ITS–90 is that another lab in another part of the world will measure the very same temperature with ease due to the advantages of a comprehensive international calibration standard featuring many conveniently spaced, reproducible, defining points spanning a wide range of temperatures.

Although "International Temperature Scale of 1990" has the word "scale" in its title, this is a misnomer that can be misleading. ITS–90 is not a scale; it is an *equipment calibration standard*. Temperatures measured with equipment calibrated per ITS–90 may be expressed using any temperature scale such as Celsius, Kelvin, Fahrenheit, or Rankine. For example, a temperature can be measured using equipment calibrated to the kelvin-based ITS–90 standard, and that value may then be converted to, and expressed as, a value on the Fahrenheit scale (e.g. 211.953 °F).

ITS–90 does not address the highly specialized equipment and procedures used for measuring temperatures extremely close to absolute zero. For instance, to measure temperatures in the nanokelvin range (billionths of a kelvin), scientists using optical lattice laser equipment to adiabatically cool atoms, turn off the entrapment lasers and simply measure how far the atoms drift over time to measure their temperature. A cesium atom with a velocity of 7 mm/s is equivalent to a temperature of about 700 nK (which was a record cold temperature achieved by the NIST in 1994).

Estimates of the differences between thermodynamic temperature and the ITS-90 (T-T_{90}) were published in 2010. It had become apparent that ITS-90 deviated considerably from PLTS-2000 in the overlapping range of 0.65 K to 2 K. To address this, a new ^{3}He vapor pressure scale was adopted, known as PTB-2006. For higher temperatures, expected values for T-T_{90} are below 0.1 mK for temperatures 4.2 K – 8 K, up to 8 mK at temperatures close to 130 K, at 0 K (by definition) at the triple point of water (273.16 °C), but rising again to 10 mK at temperatures close to 430 K, and reaching 46 mK at temperatures close to 1150 K.

Standard Interpolating Thermometers and their Ranges

Lower (K)	Upper (K)	Variations	Thermometer	Calibration and interpolation strategy
0.65	3.2	1	Helium-3 vapor pressure thermometer	Vapor pressure–temperature relationship fixed by a specified function.
1.25	2.1768	1	Helium-4 vapor pressure thermometer	Vapor pressure–temperature relationship fixed by a specified function.
2.1768	5.0	1	Helium-4 vapor pressure thermometer	Vapor pressure–temperature relationship fixed by a specified function.
3.0	24.5561	1	Helium gas thermometer	Calibrated at three fixed points in this range and interpolated in a specified way.
13.8033	1234.93	11	Platinum resistance thermometer	Resistance calibrated at various fixed points and interpolated in a specified way. Eleven distinct calibration procedures are specified.
1234.93	∞	3	Optical pyrometer	Calibrated at one fixed point, and extrapolated according to Planck's law. May be calibrated at Ag, Au, or Cu freezing point.

Defining Points

The table below lists the defining fixed points of ITS-90.

Substance and its state	Defining point (range)			
	K	**°C**	**°R**	**°F**
Triple point of hydrogen	13.8033	−259.3467	24.8459	−434.8241
Triple point of neon	24.5561	−248.5939	44.2010	−415.4690
Triple point of oxygen	54.3584	−218.7916	97.8451	−361.8249
Triple point of argon	83.8058	−189.3442	150.8504	−308.8196
Triple point of mercury	234.3156	−38.8344	421.7681	−37.9019
Triple point of water	273.16	0.01	491.69	32.02
Melting point of gallium	302.9146	29.7646	545.2463	85.5763
Freezing point of indium	429.7485	156.5985	773.5473	313.8773
Freezing point of tin	505.078	231.928	909.140	449.470
Freezing point of zinc	692.677	419.527	1,246.819	787.149
Freezing point of aluminum	933.473	660.323	1,680.251	1,220.581
Freezing point of silver	1,234.93	961.78	2,222.87	1,763.20
Freezing point of gold	1,337.33	1,064.18	2,407.19	1,947.52
Freezing point of copper	1,357.77	1,084.62	2,443.99	1,984.32

- *The triple point of water is frequently approximated by the using the melting point of water at standard conditions for temperature and pressure.*

- *Melting and freezing points are distinguished by whether heat is entering or leaving the sample when its temperature is measured.*

Thermocouple

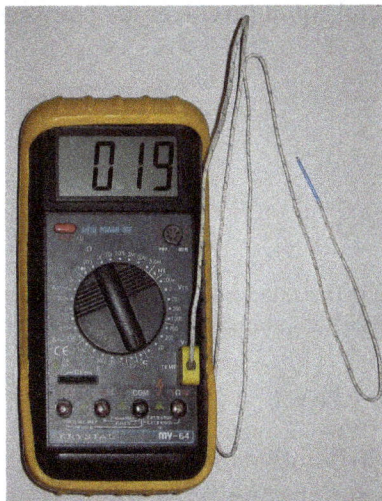

Thermocouple connected to a multimeter displaying room temperature in °C

A thermocouple is an electrical device consisting of two dissimilar conductors forming electrical junctions at differing temperatures. A thermocouple produces a temperature-dependent voltage as a result of the thermoelectric effect, and this voltage can be interpreted to measure temperature. Thermocouples are a widely used type of temperature sensor.

Commercial thermocouples are inexpensive, interchangeable, are supplied with standard connectors, and can measure a wide range of temperatures. In contrast to most other methods of temperature measurement, thermocouples are self powered and require no external form of excitation. The main limitation with thermocouples is accuracy; system errors of less than one degree Celsius (°C) can be difficult to achieve.

Thermocouples are widely used in science and industry. Applications include temperature measurement for kilns, gas turbine exhaust, diesel engines, and other industrial processes. Thermocouples are also used in homes, offices and businesses as the temperature sensors in thermostats, and also as flame sensors in safety devices for gas-powered major appliances.

Principle of Operation

In 1821, the German physicist Thomas Johann Seebeck discovered that when different metals are joined at the ends and there is a temperature difference between the joints, a magnetic field is observed. At the time Seebeck referred to this as thermo-magnetism. The magnetic field he observed was later shown to be due to thermo-electric current. In practical use, the voltage generated at a single junction of two different types of wire is what is of interest as this can be used to measure temperature at very high and low temperatures. The magnitude of the voltage depends on the types of wire used. Generally, the voltage is in the microvolt range and care must be taken to obtain a usable measurement. Although very little current flows, power can be generated by a single thermocouple junction. Power generation using multiple thermocouples, as in a thermopile, is common.

The standard configuration for thermocouple usage is shown in the figure. Briefly, the desired temperature T_{sense} is obtained using three inputs—the characteristic function $E(T)$ of the thermocouple, the measured voltage V, and the reference junctions' temperature T_{ref}. The solution to the equation $E(T_{sense}) = V + E(T_{ref})$ yields T_{sense}. These details are often hidden from the user since the reference junction block (with T_{ref} thermometer), voltmeter, and equation solver are combined into a single product.

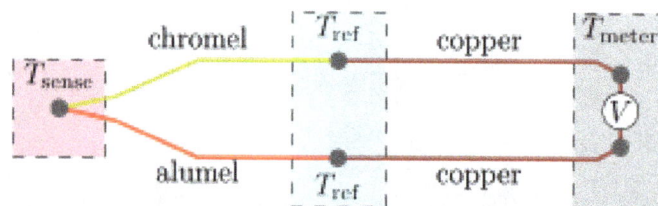

K-type thermocouple (chromel–alumel) in the standard thermocouple measurement configuration. The measured voltage V can be used to calculate temperature T_{sense}, provided that temperature T_{ref} is known.

Physical Principle: Seebeck Effect

The Seebeck effect refers to an electromotive force whenever there is a temperature gradient in a conductive material. Under open-circuit conditions where there is no internal current flow, the gradient of voltage (∇V) is directly proportional to the gradient in temperature (∇T):

$$\nabla V = -S(T)\nabla T,$$

where $S(T)$ is a temperature-dependent material property known as the Seebeck coefficient.

The standard measurement configuration shown in the figure, shows four temperature regions and thus four voltage contributions:

1. Change from T_{meter} to T_{ref}, in the lower copper wire.

2. Change from T_{ref} to T_{sense}, in the alumel wire.

3. Change from T_{sense} to T_{ref}, in the chromel wire.

4. Change from T_{ref} to T_{meter}, in the upper copper wire.

The first and fourth contributions cancel out exactly, because these regions involve the same temperature change and an identical material. As a result, T_{meter} does not influence the measured voltage. The second and third contributions do not cancel, as they involve different materials.

The measured voltage turns out to be

$$V = \int_{T_{ref}}^{T_{sense}} \left(S_+(T) - S_-(T) \right) dT,$$

where S_+ and S_- are the Seebeck coefficients of the conductors attached to the positive and negative terminals of the voltmeter, respectively (chromel and alumel in the figure).

Characteristic Function

An integral does not need to be performed for every temperature measurement. Rather, the thermocouple's behaviour is captured by a characteristic function $E(T)$, which needs only to be consulted at two arguments:

$$V = E(T_{sense}) - E(T_{ref}).$$

In terms of the Seebeck coefficients, the characteristic function is defined by

$$E(T) = \int^T S_+(T') - S_-(T') dT' + \text{const}$$

The constant of integration in this indefinite integral has no significance, but is conventionally chosen such that $E(0\,^\circ\mathrm{C}) = 0$.

Thermocouple manufacturers and metrology standards organizations such as NIST provide tables of the function $E(T)$ that have been measured and interpolated over a range of temperatures, for particular thermocouple types.

Requirement for a Reference Junction

To obtain the desired measurement of T_{sense}, it is not sufficient to just measure V. The temperature at the reference junctions T_{ref} must be already known. Two strategies are often used here:

- "Ice bath" method: The reference junction block is immersed in a semi-frozen bath of distilled water at atmospheric pressure. The precise temperature of the melting point phase transition acts as a natural thermostat, fixing T_{ref} to 0°C.

Reference junction block inside a Fluke CNX t3000 temperature meter.
Two white wires connect to a thermistor (embedded in white thermal compound)
to measure the reference junctions' temperature.

- Reference junction sensor (known as "cold junction compensation"): The reference junction block is allowed to vary in temperature, but the temperature is measured at this block using a separate temperature sensor. This secondary measurement is used to compensate for temperature variation at the junction block. The thermocouple junction is often exposed to extreme environments, while the reference junction is often mounted near the instrument's location. Semiconductor thermometer devices are often used in modern thermocouple instruments.

In both cases the value $V + E(T_{ref})$ is calculated, then the function $E(T)$ is searched for a matching value. The argument where this match occurs is the value of T_{sense}.

- A less common strategy is to use a hot reference junction. Historically, this was commonly found in temperature-critical processing plants where large numbers, often in the hundreds, of thermocouples were installed. It permitted the wiring from the field to the instrumentation or control room to be done using copper cable. Temperature control of the hot reference was either by an electrically heated, precision RTD controlled system or occasionally by a bimetallic controlled steam heater (in hazardous areas).

Practical Concerns

Thermocouples ideally should be very simple measurement devices, with each type being characterized by a precise $E(T)$ curve, independent of any other details. In reality, thermocouples are affected by issues such as alloy manufacturing uncertainties, aging effects, and circuit design mistakes/misunderstandings.

Circuit Construction

A common error in thermocouple construction is related to cold junction compensation. If an error is made on the estimation of T_{ref}, the same error will be carried over to the temperature measurement. For the simplest measurements, thermocouple wires are connected to copper far away from the hot or cold point whose temperature is measured; the cold junction is then assumed to be, at room temperature, but that temperature can vary.

Junctions should be made in a reliable manner, but there are many possible approaches to accomplish this. For low temperatures, junctions can be brazed or soldered, however it may be difficult to find a suitable flux and this may not be suitable at the sensing junction due to the solder's low melting point. Reference and extension junctions are therefore usually made with screw terminal blocks. For high temperatures, a common approach is a spot weld or crimp using a durable material. A common myth regarding thermocouples is that junctions must be made cleanly without involving a third metal, to avoid unwanted added emfs. This may result from another common misunderstanding that the voltage is generated at the junction. In fact, the junctions should in principle have uniform internal temperature, therefore no voltage is generated at the junction. The voltage is generated in the thermal gradient, along the wire.

A thermocouple produces small signals, often microvolts in magnitude. Precise measurements of this signal require an amplifier with low input offset voltage and with care taken to avoid thermal emfs from self-heating within the voltmeter itself. If the thermocouple wire has a high resistance for some reason (poor contact at junctions, or very thin wires used for fast thermal response), the measuring instrument should have high input impedance to prevent an offset in the measured voltage. A useful feature in thermocouple instrumentation will simultaneously measure resistance and detect faulty connections in the wiring or at thermocouple junctions.

Metallurgical Grades

While a thermocouple wire type is often described by its chemical composition, the actual aim is to produce a pair of wires that follow a standardized $E(T)$ curve.

Impurities affect each batch of metal differently, producing variable Seebeck coefficients. To match the standard behaviour, thermocouple wire manufacturers will deliberately mix in additional impurities to "dope" the alloy, compensating for uncontrolled variations in source material. As a result, there are standard and specialized grades of thermocouple wire, depending on the level of precision demanded in the thermocouple behaviour. Precision grades may only be available in matched pairs, where one wire is modified to compensate for deficiencies in the other wire.

A special case of thermocouple wire is known as "extension grade", designed to carry the thermoelectric circuit over a longer distance. Extension wires follow the stated $E(T)$ curve but for various reasons they are not designed to be used in extreme environments and so they cannot be used at the sensing junction in some applications. For example, an extension wire may be in a different form, such as highly flexible with stranded construction and plastic insulation, or be part of a multi-wire cable for carrying many thermocouple circuits. With expensive noble metal thermocouples, the extension wires may even be made of a completely different, cheaper material that mimics the standard type over a reduced temperature range.

Aging of Thermocouples

Thermocouples are often used at high temperatures and in reactive furnace atmospheres. In this case, the practical lifetime is limited by thermocouple aging. The thermoelectric coefficients of the wires in a thermocouple that is used to measure very high temperatures may change with time, and the measurement voltage accordingly drops. The simple relationship between the temperature difference of the junctions and the measurement voltage is only correct if each wire is homogeneous (uniform in composition). As thermocouples age in a process, their conductors can lose homogeneity due to chemical and metallurgical changes caused by extreme or prolonged exposure to high temperatures. If the aged section of the thermocouple circuit is exposed to a temperature gradient, the measured voltage will differ, resulting in error.

Aged thermocouples are only partly modified, for example being unaffected in the parts outside the furnace. For this reason, aged thermocouples cannot be taken out of their installed location and recalibrated in a bath or test furnace to determine error. This also explains why error can sometimes be observed when an aged thermocouple is pulled partly out of a furnace—as the sensor is pulled back, aged sections may see exposure to increased temperature gradients from hot to cold as the aged section now passes through the cooler refractory area, contributing significant error to the measurement. Likewise, an aged thermocouple that is pushed deeper into the furnace might sometimes provide a more accurate reading if being pushed further into the furnace causes the temperature gradient to occur only in a fresh section.

Types

Certain combinations of alloys have become popular as industry standards. Selection of the combination is driven by cost, availability, convenience, melting point, chemical properties, stability, and output. Different types are best suited for different applications. They are usually selected on the basis of the temperature range and sensitivity needed. Thermocouples with low sensitivities (B, R, and S types) have correspondingly lower resolutions. Other selection criteria include the chemical inertness of the thermocouple material and whether it is magnetic or not. Standard thermocouple types are listed below with the positive electrode (assuming $T_{\text{sense}} > T_{\text{ref}}$) first, followed by the negative electrode.

Nickel-alloy Thermocouples

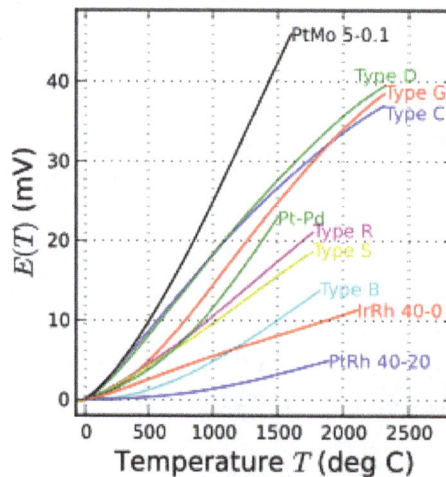

Characteristic functions for thermocouples that reach intermediate temperatures, as covered by nickel-alloy thermocouple types E, J, K, M, N, T. Also shown are the noble-metal alloy type P and the pure noble-metal combinations gold–platinum and platinum–palladium.

Type E

Type E (chromel–constantan) has a high output (68 µV/°C), which makes it well suited to cryogenic use. Additionally, it is non-magnetic. Wide range is −50°C to +740°C and narrow range is −110°C to +140°C.

Type J

Type J (iron–constantan) has a more restricted range (−40°C to +750°C) than type K but higher sensitivity of about 50 µV/°C. The Curie point of the iron (770°C) causes a smooth change in the characteristic, which determines the upper temperature limit.

Type K

Type K (chromel–alumel) is the most common general-purpose thermocouple with a sensitivity of approximately 41 µV/°C. It is inexpensive, and a wide variety of probes are available in its −200°C to +1350°C (−330°F to +2460°F) range. Type K was specified at a time when metallurgy was less advanced than it is today, and consequently characteristics may vary considerably between samples. One of the constituent metals, nickel, is magnetic; a characteristic of thermocouples made with magnetic material is that they undergo a deviation in output when the material reaches its Curie point, which occurs for type K thermocouples at around 185°C.

They operate very well in oxidizing atmospheres. If, however, a mostly reducing atmosphere (such as hydrogen with a small amount of oxygen) comes into contact with the wires, the chromium in the chromel alloy oxidizes. This reduces the emf output, and the thermocouple reads low. This phenomenon is known as *green rot*, due to the color of the affected alloy. Although not always distinctively green, the chromel wire will develop a mottled silvery skin and become magnetic. An easy way to check for this problem is to see whether the two wires are magnetic (normally, chromel is non-magnetic).

Hydrogen in the atmosphere is the usual cause of green rot. At high temperatures, it can diffuse through solid metals or an intact metal thermowell. Even a sheath of magnesium oxide insulating the thermocouple will not keep the hydrogen out.

Type M

Type M (82%Ni/18%Mo–99.2%Ni/0.8%Co, by weight) are used in vacuum furnaces for the same reasons as with type C (described below). Upper temperature is limited to 1400°C. It is less commonly used than other types.

Type N

Type N (Nicrosil–Nisil) thermocouples are suitable for use between −270°C and +1300°C, owing to its stability and oxidation resistance. Sensitivity is about 39 µV/°C at 900°C, slightly lower compared to type K.

Designed at the Defence Science and Technology Organisation (DSTO) of Australia, by Noel A. Burley, type-N thermocouples overcome the three principal characteristic types and causes of thermoelectric instability in the standard base-metal thermoelement materials:

1. A gradual and generally cumulative drift in thermal EMF on long exposure at elevated temperatures. This is observed in all base-metal thermoelement materials and is mainly due

to compositional changes caused by oxidation, carburization, or neutron irradiation that can produce transmutation in nuclear reactor environments. In the case of type-K thermocouples, manganese and aluminium atoms from the KN (negative) wire migrate to the KP (positive) wire, resulting in a down-scale drift due to chemical contamination. This effect is cumulative and irreversible.

2. A short-term cyclic change in thermal EMF on heating in the temperature range about 250–650°C, which occurs in thermocouples of types K, J, T, and E. This kind of EMF instability is associated with structural changes such as magnetic short-range order in the metallurgical composition.

3. A time-independent perturbation in thermal EMF in specific temperature ranges. This is due to composition-dependent magnetic transformations that perturb the thermal EMFs in type-K thermocouples in the range about 25–225°C, and in type J above 730°C.

The Nicrosil and Nisil thermocouple alloys show greatly enhanced thermoelectric stability relative to the other standard base-metal thermocouple alloys because their compositions substantially reduce the thermoelectric instabilities described above. This is achieved primarily by increasing component solute concentrations (chromium and silicon) in a base of nickel above those required to cause a transition from internal to external modes of oxidation, and by selecting solutes (silicon and magnesium) that preferentially oxidize to form a diffusion-barrier, and hence oxidation-inhibiting films.

Type T

Type T (copper–constantan) thermocouples are suited for measurements in the −200 to 350°C range. Often used as a differential measurement, since only copper wire touches the probes. Since both conductors are non-magnetic, there is no Curie point and thus no abrupt change in characteristics. Type-T thermocouples have a sensitivity of about 43 μV/°C. Note that copper has a much higher thermal conductivity than the alloys generally used in thermocouple constructions, and so it is necessary to exercise extra care with thermally anchoring type-T thermocouples.

Platinum/rhodium-alloy Thermocouples

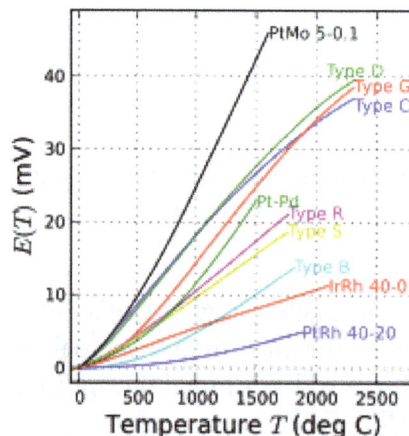

Characteristic functions for high-temperature thermocouple types, showing Pt/Rh, W/Re, Pt/Mo, and Ir/Rh-alloy thermocouples. Also shown is the Pt–Pd pure-metal thermocouple.

Types B, R, and S thermocouples use platinum or a platinum/rhodium alloy for each conductor. These are among the most stable thermocouples, but have lower sensitivity than other types, approximately 10 μV/°C. Type B, R, and S thermocouples are usually used only for high-temperature measurements due to their high cost and low sensitivity.

Type B

Type B (70%Pt/30%Rh–94%Pt/6%Rh, by weight) thermocouples are suited for use at up to 1800°C. Type-B thermocouples produce the same output at 0°C and 42°C, limiting their use below about 50°C. The emf function has a minimum around 21°C, meaning that cold-junction compensation is easily performed, since the compensation voltage is essentially a constant for a reference at typical room temperatures.

Type R

Type R (87%Pt/13%Rh–Pt, by weight) thermocouples are used up to 1600°C.

Type S

Type S (90%Pt/10%Rh–Pt, by weight) thermocouples, similar to type R, are used up to 1600°C. Before the introduction of the International Temperature Scale of 1990 (ITS-90), precision type-S thermocouples were used as the practical standard thermometers for the range of 630°C to 1064°C, based on an interpolation between the freezing points of antimony, silver, and gold. Starting with ITS-90, platinum resistance thermometers have taken over this range as standard thermometers.

Tungsten/rhenium-alloy Thermocouples

These thermocouples are well suited for measuring extremely high temperatures. Typical uses are hydrogen and inert atmospheres, as well as vacuum furnaces. They are not used in oxidizing environments at high temperatures because of embrittlement. A typical range is 0 to 2315°C, which can be extended to 2760°C in inert atmosphere and to 3000°C for brief measurements.

Type C

(95%W/5%Re–74%W/26%Re, by weight)

Type D

(97%W/3%Re–75%W/25%Re, by weight)

Type G

(W–74%W/26%Re, by weight)

Others

Chromel–gold/iron-alloy Thermocouples

In these thermocouples (chromel–gold/iron alloy), the negative wire is gold with a small fraction (0.03–0.15 atom percent) of iron. The impure gold wire gives the thermocouple a high sensitivity at low temperatures (compared to other thermocouples at that temperature), whereas the chromel wire maintains the sensitivity near room temperature. It can be used for cryogenic applications (1.2–300 K and even up to 600 K). Both the sensitivity and the temperature range depend on the iron concentration. The sensitivity is typically around 15 µV/K at low temperatures, and the lowest usable temperature varies between 1.2 and 4.2 K.

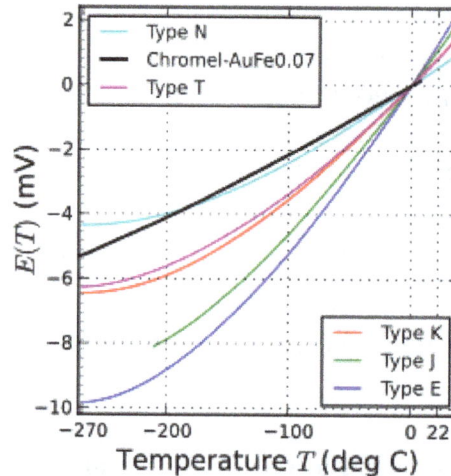

Thermocouple characteristics at low temperatures. The AuFe-based thermocouple shows a steady sensitivity down to low temperatures, whereas conventional types soon flatten out and lose sensitivity at low temperature.

Type P (Noble-metal Alloy)

Type P (55%Pd/31%Pt/14%Au–65%Au/35%Pd, by weight) thermocouples give a thermoelectric voltage that mimics the type K over the range 500°C to 1400°C, however they are constructed purely of noble metals and so shows enhanced corrosion resistance. This combination is also known as Platinel II.

Platinum/molybdenum-alloy Thermocouples

Thermocouples of platinum/molybdenum-alloy (95%Pt/5%Mo–99.9%Pt/0.1%Mo, by weight) are sometimes used in nuclear reactors, since they show a low drift from nuclear transmutation induced by neutron irradiation, compared to the platinum/rhodium-alloy types.

Iridium/rhodium Alloy Thermocouples

The use of two wires of iridium/rhodium alloys can provide a thermocouple that can be used up to about 2000°C in inert atmospheres.

Pure Noble-metal Thermocouples Au–Pt, Pt–Pd

Thermocouples made from two different, high-purity noble metals can show high accuracy even when uncalibrated, as well as low levels of drift. Two combinations in use are gold–platinum and platinum–palladium. Their main limitations are the low melting points of the metals involved (1064°C for gold and 1555°C for palladium). These thermocouples tend to be more accurate than type S, and due to their economy and simplicity are even regarded as competitive alternatives to the platinum resistance thermometers that are normally used as standard thermometers.

Skutterudite Thermocouples

NASA is developing a Multi-Mission Radioisotope Thermoelectric Generator in which the thermocouples would be made of skutterudite, which can function with a smaller temperature difference than the current tellurium designs. This would mean that an otherwise similar RTG would generate 25% more power at the beginning of a mission and at least 50% more after seventeen years. NASA hopes to use the design on the next New Frontiers mission.

Comparison of Types

The table below describes properties of several different thermocouple types. Within the tolerance columns, T represents the temperature of the hot junction, in degrees Celsius. For example, a thermocouple with a tolerance of $\pm 0.0025 \times T$ would have a tolerance of $\pm 2.5°C$ at $1000°C$.

Type	Temperature range (°C)				Tolerance class (°C)	
	Continuous		Short-term		One	Two
	Low	High	Low	High		
K	0	+1100	−180	+1300	−40 − 375: ±1.5 375 − 1000: ±0.004×T	−40 − 333: ±2.5 333 − 1200: ±0.0075×T
J	0	+750	−180	+800	−40 − 375: ±1.5 375 − 750: ±0.004×T	−40 − 333: ±2.5 333 − 750: ±0.0075×T
N	0	+1100	−270	+1300	−40 − 375: ±1.5 375 − 1000: ±0.004×T	−40 − 333: ±2.5 333 − 1200: ±0.0075×T
R	0	+1600	−50	+1700	0 − 1100: ±1.0 1100 − 1600: ±0.003×$(T−767)$	0 − 600: ±1.5 600 − 1600: ±0.0025×T
S	0	+1600	−50	+1750	0 − 1100: ±1.0 1100 − 1600: ±0.003×$(T−767)$	0 − 600: ±1.5 600 − 1600: ±0.0025×T
B	+200	+1700	0	+1820	Not available	600 − 1700: ±0.0025×T
T	−185	+300	−250	+400	−40 − 125: ±0.5 125 − 350: ±0.004×T	−40 − 133: ±1.0 133 − 350: ±0.0075×T
E	0	+800	−40	+900	−40 − 375: ±1.5 375 − 800: ±0.004×T	−40 − 333: ±2.5 333 − 900: ±0.0075×T
Chromel/ AuFe	−272	+300	N/A	N/A	Reproducibility 0.2% of the voltage. Each sensor needs individual calibration.	

Thermocouple Insulation

The wires that make up the thermocouple must be insulated from each other everywhere, except at the sensing junction. Any additional electrical contact between the wires, or contact of a wire to other conductive objects, can modify the voltage and give a false reading of temperature.

Plastics are suitable insulators for low temperatures parts of a thermocouple, whereas ceramic insulation can be used up to around 1000°C. Other concerns (abrasion and chemical resistance) also affect the suitability of materials.

Typical low cost type K thermocouple (with standard type K connector). While the wires can survive and function at high temperatures, the plastic insulation will start to break down at 300°C.

When wire insulation disintegrates, it can result in an unintended electrical contact at a different location from the desired sensing point. If such a damaged thermocouple is used in the closed loop control of a thermostat or other temperature controller, this can lead to a runaway overheating event and possibly severe damage, as the false temperature reading will typically be lower than the sensing junction temperature. Failed insulation will also typically outgas, which can lead to process contamination. For parts of thermocouples used at very high temperatures or in contamination-sensitive applications, the only suitable insulation may be vacuum or inert gas; the mechanical rigidity of the thermocouple wires is used to keep them separated.

Table of Insulation Materials

Type of Insulation	Max. continuous temperature	Max. single reading	Abrasion resistance	Moisture resistance	Chemical resistance
Mica–glass tape	649°C/1200°F	705°C/1300°F	Good	Fair	Good
TFE tape, TFE–glass tape	649°C/1200°F	705°C/1300°F	Good	Fair	Good
Vitreous-silica braid	871°C/1600°F	1093°C/2000°F	Fair	Poor	Poor
Double glass braid	482°C/900°F	538°C/1000°F	Good	Good	Good
Enamel–glass braid	482°C /900°F	538°C/1000°F	Fair	Good	Good
Double glass wrap	482°C/900°F	427°C/800°F	Fair	Good	Good
Non-impregnated glass braid	482°C/900°F	427°C/800°F	Poor	Poor	Fair
Skive TFE tape, TFE–glass braid	482°C/900°F	538°C/1000°F	Good	Excellent	Excellent
Double cotton braid	88°C/190°F	120°C/248°F	Good	Good	Poor
"S" glass with binder	704°C/1300°F	871°C/1600°F	Fair	Fair	Good
Nextel ceramic fiber	1204°C/2200°F	1427°C/2600°F	Fair	Fair	Fair
Polyvinyl/nylon	105°C/221°F	120°C/248°F	Excellent	Excellent	Good
Polyvinyl	105°C/221°F	105°C/221°F	Good	Excellent	Good
Nylon	150°C/302°F	130°C/266°F	Excellent	Good	Good
PVC	105°C/221°F	105°C/221°F	Good	Excellent	Good
FEP	204°C/400°F	260°C/500°F	Excellent	Excellent	Excellent

Type of Insulation	Max. continuous temperature	Max. single reading	Abrasion resistance	Moisture resistance	Chemical resistance
Wrapped and fused TFE	260°C/500°F	316°C/600°F	Good	Excellent	Excellent
Kapton	316°C/600°F	427°C/800°F	Excellent	Excellent	Excellent
Tefzel	150°C/302°F	200°C/392°F	Excellent	Excellent	Excellent
PFA	260°C/500°F	290°C/550°F	Excellent	Excellent	Excellent
T300*	300°C	–	Good	Excellent	Excellent

Temperature ratings for insulations may vary based on what the overall thermocouple construction cable consists of.

Note: T300 is a new high-temperature material that was recently approved by UL for 300°C operating temperatures.

Applications

Thermocouples are suitable for measuring over a large temperature range, from –270 up to 3000°C (for a short time, in inert atmosphere). Applications include temperature measurement for kilns, gas turbine exhaust, diesel engines, other industrial processes and fog machines. They are less suitable for applications where smaller temperature differences need to be measured with high accuracy, for example the range 0–100°C with 0.1°C accuracy. For such applications thermistors, silicon bandgap temperature sensors and resistance thermometers are more suitable.

Steel Industry

Type B, S, R and K thermocouples are used extensively in the steel and iron industries to monitor temperatures and chemistry throughout the steel making process. Disposable, immersible, type S thermocouples are regularly used in the electric arc furnace process to accurately measure the temperature of steel before tapping. The cooling curve of a small steel sample can be analyzed and used to estimate the carbon content of molten steel.

Gas Appliance Safety

A thermocouple (the right most tube) inside the burner assembly of a water heater

Thermocouple connection in gas appliances. The end ball (contact) on the left is insulated from the fitting by an insulating washer. The thermocouple line consists of copper wire, insulator and outer metal
(usually copper) sheath which is also used as ground.

Many gas-fed heating appliances such as ovens and water heaters make use of a pilot flame to ignite the main gas burner when required. If the pilot flame goes out, unburned gas may be released, which is an explosion risk and a health hazard. To prevent this, some appliances use a thermocouple in a fail-safe circuit to sense when the pilot light is burning. The tip of the thermocouple is placed in the pilot flame, generating a voltage which operates the supply valve which feeds gas to the pilot. So long as the pilot flame remains lit, the thermocouple remains hot, and the pilot gas valve is held open. If the pilot light goes out, the thermocouple temperature falls, causing the voltage across the thermocouple to drop and the valve to close.

Some combined main burner and pilot gas valves (mainly by Honeywell) reduce the power demand to within the range of a single universal thermocouple heated by a pilot (25 mV open circuit falling by half with the coil connected to a 10–12 mV, 0.2–0.25 A source, typically) by sizing the coil to be able to hold the valve open against a light spring, but only after the initial turning-on force is provided by the user pressing and holding a knob to compress the spring during lighting of the pilot. These systems are identifiable by the "press and hold for x minutes" in the pilot lighting instructions. (The holding current requirement of such a valve is much less than a bigger solenoid designed for pulling the valve in from a closed position would require.) Special test sets are made to confirm the valve let-go and holding currents, because an ordinary milliammeter cannot be used as it introduces more resistance than the gas valve coil. Apart from testing the open circuit voltage of the thermocouple, and the near short-circuit DC continuity through the thermocouple gas valve coil, the easiest non-specialist test is substitution of a known good gas valve.

Some systems, known as millivolt control systems, extend the thermocouple concept to both open and close the main gas valve as well. Not only does the voltage created by the pilot thermocouple activate the pilot gas valve, it is also routed through a thermostat to power the main gas valve as well. Here, a larger voltage is needed than in a pilot flame safety system described above, and a thermopile is used rather than a single thermocouple. Such a system requires no external source of electricity for its operation and thus can operate during a power failure, provided that all the other related system components allow for this. This excludes common forced air furnaces because external electrical power is required to operate the blower motor, but this feature is especially useful for un-powered convection heaters. A similar gas shut-off safety mechanism using a thermocouple is sometimes employed to ensure that the main burner ignites within a certain time period, shutting off the main burner gas supply valve should that not happen.

Out of concern about energy wasted by the standing pilot flame, designers of many newer appliances have switched to an electronically controlled pilot-less ignition, also called intermittent ignition. With no standing pilot flame, there is no risk of gas buildup should the flame go out, so these appliances do not need thermocouple-based pilot safety switches. As these designs lose the benefit of operation without a continuous source of electricity, standing pilots are still used in some appliances. The exception is later model instantaneous (aka "tankless") water heaters that use the flow of water to generate the current required to ignite the gas burner; these designs also use a thermocouple as a safety cut-off device in the event the gas fails to ignite, or if the flame is extinguished.

Thermopile Radiation Sensors

Thermopiles are used for measuring the intensity of incident radiation, typically visible or infrared light, which heats the hot junctions, while the cold junctions are on a heat sink. It is possible to measure radiative intensities of only a few $\mu W/cm^2$ with commercially available thermopile sensors. For example, some laser power meters are based on such sensors; these are specifically known as thermopile laser sensor.

The principle of operation of a thermopile sensor is distinct from that of a bolometer, as the latter relies on a change in resistance.

Manufacturing

Thermocouples can generally be used in the testing of prototype electrical and mechanical apparatus. For example, switchgear under test for its current carrying capacity may have thermocouples installed and monitored during a heat run test, to confirm that the temperature rise at rated current does not exceed designed limits.

Power Production

A thermocouple can produce current to drive some processes directly, without the need for extra circuitry and power sources. For example, the power from a thermocouple can activate a valve when a temperature difference arises. The electrical energy generated by a thermocouple is converted from the heat which must be supplied to the hot side to maintain the electric potential. A continuous transfer of heat is necessary because the current flowing through the thermocouple tends to cause the hot side to cool down and the cold side to heat up (the Peltier effect).

Thermocouples can be connected in series to form a thermopile, where all the hot junctions are exposed to a higher temperature and all the cold junctions to a lower temperature. The output is the sum of the voltages across the individual junctions, giving larger voltage and power output. In a radioisotope thermoelectric generator, the radioactive decay of transuranic elements as a heat source has been used to power spacecraft on missions too far from the Sun to use solar power.

Thermopiles heated by kerosene lamps were used to run batteryless radio receivers in isolated areas. There are commercially produced lanterns that use the heat from a candle to run several light-emitting diodes, and thermoelectrically-powered fans to improve air circulation and heat distribution in wood stoves.

Process Plants

Chemical production and petroleum refineries will usually employ computers for logging and for limit testing the many temperatures associated with a process, typically numbering in the hundreds. For such cases, a number of thermocouple leads will be brought to a common reference block (a large block of copper) containing the second thermocouple of each circuit. The temperature of the block is in turn measured by a thermistor. Simple computations are used to determine the temperature at each measured location.

Thermocouple as Vacuum Gauge

A thermocouple can be used as a vacuum gauge over the range of approximately 0.001 to 1 torr absolute pressure. In this pressure range, the mean free path of the gas is comparable to the dimensions of the vacuum chamber, and the flow regime is neither purely viscous nor purely molecular. In this configuration, the thermocouple junction is attached to the centre of a short heating wire, which is usually energised by a constant current of about 5 mA, and the heat is removed at a rate related to the thermal conductivity of the gas. It may be possible to superimpose AC heating on the thermocouple directly, making the sensor a 2-wire device, but those on the market appear to all be 4-wire devices, with separate terminals for the heater and the thermocouple.

The temperature detected at the thermocouple junction depends on the thermal conductivity of the surrounding gas, which depends on the pressure of the gas. The potential difference measured by a thermocouple is proportional to the square of pressure over the low- to medium-vacuum range. At higher (viscous flow) and lower (molecular flow) pressures, the thermal conductivity of air or any other gas is essentially independent of pressure. The thermocouple was first used as a vacuum gauge by Voege in 1906. The mathematical model for the thermocouple as a vacuum gauge is quite complicated, as explained in detail by Van Atta, but can be simplified to:

$$P = \frac{B(V^2 - V_0^2)}{V_0^2},$$

where P is the gas pressure, B is a constant that depends on the thermocouple temperature, the gas composition and the vacuum-chamber geometry, V_0 is the thermocouple voltage at zero pressure (absolute), and V is the voltage indicated by the thermocouple.

The alternative is the Pirani gauge, which operates in a similar way, over approximately the same pressure range, but is only a 2-terminal device, sensing the change in resistance with temperature of a thin electrically heated wire, rather than using a thermocouple.

References

- "Mise en pratique for the definition of the kelvin" (PDF). Sèvres, France: Consultative Committee for Thermometry (CCT), International Committee for Weights and Measures (CIPM). 2011. Retrieved 25 June 2013

- Giuseppe Morandi; F Napoli; E Ercolessi. Statistical mechanics : an intermediate course. Singapore ; River Edge, N.J. : World Scientific, ©2001. pp. 6~7. ISBN 978-981-02-4477-4

- Baker, Donald G. (June 1975). "Effect of Observation Time on Mean Temperature Estimation". Journal of Applied Meteorology. 14 (4): 471–476. Bibcode:1975JApMe..14..471B. doi:10.1175/1520-0450(1975)014<0471:EOOTOM>2.0.CO;2

- D. Olinger; J. Gray; R. Felice (2007-10-14). Successful Pyrometry in Investment Casting (PDF). Investment Casting Institute 55th Technical Conference and Expo,. Investment Casting Institute. Retrieved 2015-04-02

- Carl S. Helrich (2009). Modern Thermodynamics with Statistical Mechanics. Berlin, Heidelberg: Springer Berlin Heidelberg. ISBN 978-3-540-85417-3

- D. Ng & G. Fralick (2001). "Use of a multiwavelength pyrometer in several elevated temperature aerospace applications". Review Scientific Instruments. 72 (2): 1522. Bibcode:2001RScI...72.1522N. doi:10.1063/1.1340558

- Kerlin, T.W. & Johnson, M.P. (2012). Practical Thermocouple Thermometry (2nd Ed.). Research Triangle Park: ISA. pp. 110–112. ISBN 978-1-937560-27-0

Permissions

Index

www.ingramcontent.com/pod-product-compliance
Lightning Source LLC
Chambersburg PA
CBHW082015190326
41458CB00010B/3200